Modernity, Metatheory, and the Temporal-Spatial Divide

This book is about how modernity affects our perceptions of time and space. Its main argument is that geographical space is used to control temporal progress by channeling it to benefit particular political, economic and social interests, or by halting it altogether. By incorporating the ancient Greek myth of the Titanomachy as a conceptual metaphor to explore the elemental ideas of time and space, the author argues that hegemonic interests have developed spatial hierarchy into a comprehensive system of technocratic monoculture, which interrupts temporal development in order to maintain exclusive power and authority. This spatial stasis is reinforced through the control of historical narratives and geographical settings. While increasingly comprehensive, the author argues that this state of affairs can best be challenged by focusing on the development of "unmappable places" which presently exist within the sociospatial matrix of the modern world.

Michael Kimaid is an Associate Professor of History and Geography at Bowling Green State University, Firelands College.

Routledge Approaches to History

1 **Imprisoned by History**
   Aspects of Historicized Life
   Martin Davies

2 **Narrative Projections of a Black British History**
   Eva Ulrike Pirker

3 **Integrity in Historical Research**
   Edited by Tony Gibbons and Emily Sutherland

4 **History, Memory, and State-Sponsored Violence**
   Time and Justice
   Berber Bevernage

5 **Frank Ankersmit's Lost Historical Cause**
   A Journey from Language to Experience
   Peter P. Icke

6 **Popularizing National Pasts**
   1800 to the Present
   Edited by Stefan Berger, Chris Lorenz and Billie Melman

7 **The Fiction of History**
   Alexander Lyon Macfie

8 **The Rise and Propagation of Historical Professionalism**
   Rolf Torstendahl

9 **The Material of World History**
   Edited by Tina Mai Chen and David S. Churchill

10 **Modernity, Metatheory, and the Temporal-Spatial Divide**
   From *Mythos* to *Techne*
   Michael Kimaid

Modernity, Metatheory, and the Temporal-Spatial Divide
From *Mythos* to *Techne*

Michael Kimaid

LONDON AND NEW YORK

First published 2015 by Routledge

2 Park Square, Milton Park, Abingdon, Oxfordshire OX14 4RN
52 Vanderbilt Avenue, New York, NY 10017

Routledge is an imprint of the Taylor & Francis Group, an informa business

First issued in paperback 2019

Copyright © 2015 Taylor & Francis

The right of Michael Kimaid to be identified as author of this work has been asserted in accordance with sections 77 and 78 of the Copyright, Designs and Patents Act 1988.

All rights reserved. No part of this book may be reprinted or reproduced or utilized in any form or by any electronic, mechanical, or other means, now known or hereafter invented, including photocopying and recording, or in any information storage or retrieval system, without permission in writing from the publishers.

Notice:
Product or corporate names may be trademarks or registered trademarks, and are used only for identification and explanation without intent to infringe.

Library of Congress Cataloging-in-Publication Data
Kimaid, Michael.
Modernity, metatheory, and the temporal-spatial divide : from mythos to techne / by Michael Kimaid. — 1st [edition].
　　pages cm. — (Routledge approaches to history ; 10)
Includes bibliographical references and index.
　1. Space and time.　2. Civilization, Modern.　I. Title.
　BD632.K565 2015
　115—dc23
　2014045546

ISBN: 978-1-138-83261-9 (hbk)
ISBN: 978-0-367-26381-2 (pbk)

Typeset in Sabon
by Apex CoVantage, LLC

For Ethan.

Contents

List of Figures		ix
Preface		xi
	Introduction	1
1	Modernity and Its Discontents	13
2	Modern Timespace	49
3	Technocratic Monoculture	82
4	Of Spectacles and Monuments	123
	Conclusion: Unmappable Places	169
	Bibliography	189
	Index	203

Figures

I.1	A Greek Vase Depicting a Scene from the Titanomachy, c.550 BC.	2
3.1	*Science vs. Chaos* by Howard Scott, 1933.	92
3.2	*A map of the most inhabited part of Virginia containing the whole province of Maryland with part of Pensilvania, New Jersey and North Carolina.* Drawn by Joshua Fry & Peter Jefferson in 1751.	105
4.1	*The Sleep of Reason Produces Monsters* (*El sueño de la razon produce monstrous*) by Francisco Goya, 1797–1799.	126
4.2	Boston Massacre Site Memorial, 2014.	142
4.3	Tyburn Memorial.	143
4.4	Base of the Fort Meigs Memorial Obelisk, 2008.	147
4.5	Haymarket Police Monument.	148
4.6	Haymarket Martyrs Memorial.	150
4.7	Haymarket Memorial.	150
4.8	9/11 Memorial Museum Store (exterior view), 2014.	155
4.9	9/11 Memorial Museum Store (interior view), 2014.	155
4.10	Statue of Hebe in Tompkins Square Park, 2014.	163

Preface

This book began in the classroom. Teaching both History and Geography offers perspective; time and space, ancient and modern, here and there, then and now. Over the course of a few years ranging between disciplines with reckless abandon, I started to identify constants in the variables of the readings. Names, places, and dates change, but the conditions in which they manifest remain: power, oppression, injustice, and resistance. Rather than focus on a particular time and place, which others have done much more effectively than I ever could, I decided to turn my attention to theory, to try to understand how these constants persist while the variables swirl in a dizzying maelstrom around them.

I first tested the ideological waters of my thesis at the "Revisiting Modernity—A Conversation across Disciplines" conference at The University of California at Berkeley in 2009. Penelope Ismay and Arpita Roy were kind enough to include me in the proceedings, and I remain grateful for the opportunity. The next test for my ideas was the "Taking Control" conference at the University of London, organized by Luke Evans and Alexej Ulbricht in 2011. The challenge there was to merge theory and practice, to make ideas useful, and to have a response to the question that all academics face at some point in their career: "So What?"

Between then and now, I was able to work on the bulk of this text on a Visiting Research Fellowship through The Bruce Centre for American Studies at Keele University in Staffordshire, England. I owe a great deal of thanks to Axel Schaefer and Timothy Lustig for facilitating my visit there, and helping to make it as productive as it was. The interest of Oliver Harris and Leslie Powner, and the conversations that followed, have made this a better book. Along the way, I had the good fortune of striking up a friendship with Alun Munslow, whose encouragement and insight was instrumental in the direction and completion of the work. In addition, I am eternally grateful to Molly Mosher at Routledge Press for moving my proposal through the system, to my editor Max Novick for taking notice of it and seeing it through, and to Jennifer Morrow for facilitating everything with an eternal patience. Thanks also to Christina Faria for the thorough and exacting copy editing work and to the Apex Covantage prepress team for helping to see the project through from manuscript to book.

I am very thankful for the support of so many people who have contributed in their own ways to this project. Scott Martin has had a great deal to do with shaping my academic vision and voice, as have Edmund Danziger, Thomas Knox, and Robert Buffington. I feel very lucky to be able to call Peter Way both a mentor and a friend. My comrades in the Bowling Green State University Faculty Association: David Jackson, Lori Liggett, Julie Haught, James Evans, Joel O'Dorisio, Jamie Stuart, and Andy Schocket, along with many others, have helped remind me that making history is as important as studying it. At BGSU-Firelands, I am grateful for the friendship of Timothy Jurkovac, the sage advice of Chris Mruk, and the unwavering support of my department chair, Victor Odafe.

Outside of the academy, there are a great many people who have shown me how to put theory into practice: to turn ideas into real things. Chief among them are Gabriel Beam and Ryan Dohm, who I am fortunate to have the opportunity to collaborate with. Terri Miller, J. S. Makkos, Tom Orange, Lisa Miralia, Rob Galo, Isaac Hand, Mike Shiflet, Jose Luna, Curt Brown, Karl Vorndran, Keir Neuringer, Bryan Day, Colin Helb, Josh Eppert, and Jay Kreimer are among the many others who remind me what community and neighborhood really are. Steve, Dan and the entire Piotrowicz Family, Doug Hall, Karel Blaas, Chad Jakubowski, Adam Rugnetta, Steve Nosenchuck, Jeff and Stephanie Johnson, Brian Frawley, Joseph Smith, and Michael Hanmer have all been the truest of friends along the way.

My family is at the root of all of this. My parents, Ron and Sharon Kimaid, have been steadfast in their love and support. My grandparents, aunts, uncles, and cousins, and the times we have shared together, have vitally shaped my perspective of the world. Joe and Lorraine Roecklein, and Jessica and Gary Gralak have broadened that perspective. My wife, Josie, who has been with me every step of the way, and my son Ethan, to whom this book is dedicated, have both endured the process of producing this work along with me, transatlantic travel and all. This book is every bit as much theirs as it is mine.

Michael Kimaid
Perrysburg, Ohio
July 4, 2014.

Introduction

TITANOMACHY

With Hesiod, "From the Muses of Helicon, let us begin our singing."[1] (Let us, though, sing in dissonant counterpoint).[2]

In the Greek Titanomachy, before the abstract rationalization of eternal time and infinite space, Kronos escaped the depths of Tartarus where his father had imprisoned him to liberate his mother Gaea (Earth) from the abuse of Uranus with a bloody fury. In doing so, Kronos freed all of his Titanic brothers and sisters, save for the perpetually obstructive Cyclopes and Hekatonkheires, whose eventual liberation would change the course of both the revolt and human history. Kronos, time itself mythologized, had come to set his rule upon the Chasm of Chaos, or infinite open space. The ancient Greeks associated Kronos's rule with "The Golden Age," the first sequence of the "Ages of Man," in which there was no need for laws, the common good was observed, and self-interested immorality was absent. In Hesiod's words, humanity "dwelt in ease and peace."[3]

To protect this Golden Age, Kronos then tried to prevent the cycle of rebellious patricide which he had begun against his father by devouring his own offspring. His son Zeus had managed to escape the filial cannibalism of Kronos and rose to cast him out in the Titanic Wars, assuming ultimate power on Mount Olympus for himself (See Figure I.1). In the course of battle, Zeus released the long imprisoned Cyclopes and Hekatonkheires from Tartarus. The former provided thunder and lightning and the latter provided the hundred-handed strength with which Zeus' victory was complete. In the meantime, the great conflagration had spread to Chaos itself. Where Kronos went in defeat is uncertain. According to Hesiod, Kronos was cast back into the abyss of Tartarus; "put . . . down into in everlasting shade." Pausanias the Traveler suggests that he was sent far away to the west, beyond the Pillars of Hercules.[4] Wherever he went, below or beyond, Kronos was relegated to the marginal periphery. Time was no longer central to the worldly condition.

The Golden Age of commonwealth and mutuality was over. In the next age, Hesiod observes that "people could not restrain themselves from

2 *Introduction*

Figure I.1 A Greek Vase Depicting a Scene from the Titanomachy, c.550 BC.

Photographer: Bibi Saint-Pol. Staatliche Antikensammlungen via Wikimedia Commons. http://upload.wikimedia.org/wikipedia/commons/d/d9/Zeus_Typhon_Staatliche_Antikensammlungen_596.jpg.

crime against each other."[5] Throughout the successively degenerative ages of history as Hesiod recounted them, the Cyclopes and the Hekatonkheires stood watch, protecting the interests of Zeus as he wielded power from high and distant Olympus, ever scanning the Chaos below and beyond for the return of Kronos. With Zeus's reign came mortal hardship and peril. Zeus learned to make mortals dependent on his favor, as he ultimately came to control the Fates as a way of maintaining his power.[6] Meanwhile, Kronos in his exile found a renewed partnership with Ananke, the Goddess of Inevitability. History accounts for Kronos's and Ananke's occasional rebellious incursions from the peripheral abyss into Zeus's realm, but in the meantime, the sentinel Cyclopes and Hekatonkheires have multiplied their monocular gaze and centurion arms in vigilant defense of their patron's autocracy.

THE SEMIOTIC MEANING OF TITANOMACHY: TIME AND SPACE

Nearly three thousand years after Hesiod accounted for all of this in *Theogony*, the multinational corporation Power Telecomm Inc., which specializes in the development and manufacture of surveillance equipment, announced that it "proudly introduces its new line of color CCD [Charge Coupled

Device] cameras, the Cyclops series."[7] With this, the allegorical implications of the Titanomachy become evident. The Cyclopes, once *mythology*, are become manifest in *technology*. Their purpose, however, remains constant: to watch the periphery and protect the core; to protect Olympus from Tartarus; to protect order from Chaos. What Hesiod explained in mythological terms, the surveillance of space to protect institutional hierarchy from rebellion has become itself institutional. In that the ancient has become modern, we are connected by heuristic *mythos* and epistemic *techne* to the most metaphysical of power struggles: between time and space.

What does a mythological tale from the Dark Ages of ancient Greece have to do with the present, and what can we learn from its retelling? To begin to understand the connection between Hesiod's account of how the pantheon of Olympus came to be, and what it means today, the story needs to be refracted through the prism of semiotic inquiry. Semiotics helps unlock the symbolism of the ancient tale of the Titanomachy and begins to explore its implications for the modern condition. While the lore of mythology may seem as obtuse as it does fantastic, the ideology behind it is prescient. Helen Morales observes that "myth is a *process* as much as a *thing*."[8] Semiotics, a discipline that focuses on articulating the meanings of signs and symbols that constitute all forms of human communication, is a key to help unlock the fundamental meaning behind the myth as it resonates millennia after Hesiod committed the story to writing.

Semiotics provides a way to understand meaning through unpacking the symbolism inherent in all forms of human communication. Signs and codes can be verbal, nonverbal, image-related, or spatial. The Swiss linguist Ferdinand Saussere and the American philosopher Charles Sanders Pierce laid the foundation for semiotic theory in the late nineteenth and early twentieth centuries. Saussere declared that semiotics was a "science which studies the life of signs within society is conceivable. Semiology would show what constitutes signs, what laws govern them."[9] Saussere rooted the task of semiotic inquiry in psychology and linguistics. Pierce, working independent of Saussere's observations, wrote that "logic is *formal semiotic* . . . It is from this definition that I deduce the principles of logic by mathematical reasoning."[10] Taken together, semiotic inquiry suggests that there is a subtextual (or hypertextual) meaning to human communication, and that the exploration of those meanings, inherently psychological, follows a logical form of inquiry both philosophical and mathematical in its origins. Semiotics offers a key to explain what codes are present in mythological narratives, which makes them so simultaneously obtuse and resonant. What the signs and symbols of mythology signify and symbolize becomes evident through semiotic inquiry.[11]

There are six elements of Hesiod's narrative of the Titanomachy that elucidate themes of present consequence. The first is the being of Kronos, depicted as a monster by Hesiod, who has an obvious etymological relation to the concept of time. Next is Zeus, who rises up and leads the challenge against Kronos for power, ultimately winning it and establishing himself as a supreme deity on high Mount Olympus. Zeus, "in the air," and "on high," controls and commands the earth from above, establishing the principal of spatial hierarchy in the creation story of Greek mythology. Zeus and Olympus taken together as the second and third thematic elements of the story represent the personage and the place of the elite ruling class today. Hesiod wrote that Zeus was ultimately successful by freeing the Cyclopes and the Hekatonkheires from Tartarus where Kronos had imprisoned them and conscripting them for battle. The monocular Cyclopes, then, become the fourth element of the narrative to consider. These one eyed giants "whose hearts were insolent," serve in the semiotic process of myth as recognizable symbols of our present surveillance culture. During the war, they provide Zeus with the power, in the form of thunder and lightning, which helps him defeat Kronos for control of the universe. After the war with the Titans, at Zeus's behest, the Cyclopes built a wall around Tartarus where Kronos and the Titans were sent, which they guarded along with the Hekatonkheires. The Hekatonkheires, our fifth element of interest, were fifty-headed, hundred-armed giants "who loved war insatiably"; they represent the expansive institutions of the corporate capitalism today. During "The Golden Age," Kronos had checked their aggression by sending them to Tartarus. Their might and fury helped Zeus defeat Kronos in battle, and afterwards, they became Kronos's jailers. Tartarus then, is our final element of thematic consequence in the story. A place of darkness and gloom, "as far beneath the earth, as earth is far below the heavens," it represents the marginal periphery from where insurgency comes. Hesiod's conclusion of the battle reestablishes the spatially hierarchical framework of command and control which Zeus controls from high Olympus, enforced by the Cyclopes and Hekatonkheires. With the onset of modernity, it has been the preoccupation of the institutional power structure to triangulate and control the marginal periphery to eliminate the threat of insurgency. By spatially relegating Kronos, time itself "bound up in painful chains," to Tartarus, bounding it with walls and assigning the gaze of the Cyclopes and the hundred-handed strength of the Hekatonkheires to police and enforce the boundary, the hierarchical control of space to prevent a form of temporal liberation which would restore a "Golden Age" becomes the ultimate interpretative process of the Titanomachy myth.

　　The difference between the temporal and spatial realities that Kronos and Zeus represent is elemental. Historical time is predicated on and marked by degrees of transformation. Otherwise stated, the basic function of historical inquiry is to recognize and account for change. Time is variable; things are different. Over the course of history, both time and the space in which it manifests have been triangulated in order to prevent unsanctioned change from occurring; the chaos of the periphery has been surveyed, mapped, and bounded. That is to say, they have been systematically arranged for purposes of regulation, control, and the enforcement of hierarchical order. This is not to say that those measures have been entirely successful. The hegemonic control of temporal and spatial conditions is far from complete. It has, however, systematically advanced, since the onset of modernity, to a degree that is fairly comprehensive in nature. Within such conditions, finding opportunities to effectively resist the imposition of authority and control becomes increasingly difficult.

The geographical relation of space, since Zeus set loose the Cyclopes and Hekatonkheires to defend the Olympian ordering of Chaos is predicated on maintaining power through enforcing stasis, and preventing the temporal progression of history from occurring. With the peripheral challenge in check, the hierarchical order in power pursues an inherently degenerative agenda. Hesiod's account of the subsequent Ages of Man under Zeus's rule describes the regressive deterioration of the quality of life for the living. Meanwhile, Kronos and Ananke, time and inevitability mythologized, together represent the idea of temporal liberation from the increasingly oppressive conditions that the Olympian order imposes on humankind. Over the course of history, the triangulation of geographical space has often served to control, challenge, or altogether prevent the dynamics of historical change. As the triangulation of space has become more exacting with the onset of modernity and its exponentially rapid shift from *mythos* to *techne*, the dynamic of temporal change has also become increasingly difficult to assert and to realize in surveilled space. When properly surveilled and enforced, space is static; the agents of hierarchical power and control—be they economic, social, or political—prevent unregulated change from occurring. The Cyclopes and Hekatonkheires, serving as conceptual metaphors for surveillance technology and global capitalism respectively, protect high Olympus from Typhon's monsters who would unseat Zeus and return the world to Kronos and "The Golden Age." Georg Wilhelm Friedrich Hegel acknowledged the mythological role of the Cyclopes as a protector of institutional power in Greek history. In *Lectures on the Philosophy of History* (1836), he related the characterization of fortresses around royal houses in Argolis as "Cyclopian," and cited the myth of Proetus, ruler of Argos, who "brought with him from Lycia the Cyclopes who built these walls."[12]

HYPOTHESIS, THESIS, AND EXEGESIS

This book is ultimately about how the conditions of modernity affect our perceptions of time and space. By modernity, I mean the present tense, but also the social, economic, and political factors that define the present. The question of modernity is a relative one, and scholars debate the conditions of modernity, its context, where and when the term applies, whether multiple modernities or a single "modernity" exists, whether we are living beyond the conditions of modernity (in an era of postmodernity), or whether it even ever existed at all. These are all worthwhile discussions, but in the interests of moving the conversation forward, I want to acknowledge the conditions which define the present in the context of history and understand their temporal and spatial ramifications.

Modernity, then, is the third of three historical epochs within the tradition of western history. It is preceded by the ancient and the medieval, which are generally regarded in both temporal and spatial terms. Ancient history within the western tradition begins as Greece emerges from its "dark ages" in 800 BC, and has a somewhat drawn-out conclusion with the fall, decline, or demise of the Roman Empire in the fifth century AD. From there, the epochal narrative of western history moves into the medieval phase, and this temporal shift accompanies a spatial shift, from the connected Mediterranean world, which connected Africa, Asia, and Europe through economic, political, and cultural connections, to a more provincial and isolated European focus. Historians of the medieval period generally consider European history from the fifth century through the Renaissance, which represented a return to classical forms of knowledge and ideological attitudes from the fourteenth to the seventeenth centuries.

With the ancient and the medieval aspects of western history as a backdrop, the modern epoch comes into focus. From the fourteenth and fifteenth centuries onward, beginning in Europe, the historical record shows the exponential proliferation of four factors which arguably outline the parameters of modernity. The first of these four factors is technology. The growth of technology in every respect: transportation, communication, martial, and medicinal, among others, has shaped the course of modern history in significant and profound ways. While technologies have existed since prehistoric time and outside of the scope of western topography, the exponential expansion of technologies of western origin, from electricity to the airplane to the nuclear weapon, is a fundamental factor of the modern condition writ large.

The dual force of urbanization and nationalism is the second of four factors that helps to set modernity apart from the ancient and medieval. While abstract political identities and cities are not phenomena exclusive to western history or modernity, the imposition of national boundaries and the comprehensive political, economic, and social framework that exist within them is largely a product of nineteenth-century western thought and design. Similarly, the modern city, as a political, economic, and cultural entity, emerges from the smoke clouds of the Industrial Revolution in Europe and America, and is a model for urbanization around the globe. In 2009, for the first time in human history, the number of people living in urban areas surpassed the number living in rural areas. Together, nationalism and urbanization represent a fundamental shift in the organization of human populations, in both physical and ideological terms, that we can identify as fundamentally modern in design.

The expansion of a capitalist market economy is the third factor which makes modernity something distinct from the ancient and medieval pasts. Here again, this is not to say that markets did not exist prior to the fourteenth and fifteenth centuries. If anything, history shows us that "globalization" as an economic force is a process begun in antiquity. Again, it is the scope and scale of expansion that makes modernity distinct. In the ancient and medieval pasts, markets were limited by space and time. Technology, urbanization, and nationalism have all contributed to the exponential expansion of the market economy, which can theoretically be engaged anytime, anywhere in modernity. The process of commodification is elemental to the growth and expansion of capitalism, and most striking in the context of the modern market is the commodification of knowledge. "The Information Age" moves the market from the physical to the ideological, from the tangible to the abstract. In this process, thoughts and ideas become commodities, which can be bought and sold by marketers, or appropriated by the state to monitor and regulate accordingly.

The last factor that arguably defines modernity as something distinct from a historical past is the presence of a democratic ideology. It is important to note that while the diffusion of democratic theory is a historical factor in the context of defining modernity, it is not necessarily always found in practice. Here again, while western historians like to point to ancient Greece as the foundation of democracy, there is ample evidence that shows its development in multiple places at multiple times across the historical and geographical spectra. It is the proliferation of western state democracy that resonates first out of the American Revolution in the eighteenth century, and spreads geographically over time all the way to Tiananmen and Tehrir Squares in the twentieth and twenty-first centuries. Democratic ideology is historically the thing that keeps the excesses of the aforementioned factors: technology, nationalism and urbanization, and capitalist market economies, in check, and prevents them from institutionalizing inequality on political, economic, and social levels.

So then, how does this modernity, and the conditions that define it, affect our perceptions of time and space? To answer this question, I turn to metatheory, particularly the fields of teleology and ontology as modes of existential inquiry. Teleology is the argument that "final causes" exist in the sum total of human endeavor. That is to say, historical events and conditions happen "for a reason." Those final causes, and the reason that the historical events leading up to them occur, are all very much a matter of debate. This debate fundamentally affects our collective perception of time. To some, who point to the eighteenth-century historian Georg Friederich Hegel as a source, history is humanity's quest to embody "the absolute idea"; history was a progression toward an ultimate condition, a utopia. Others, who reach back further, cite Judeo-Christian theology to suggest that humanity is inherently degenerative, and that history is an eschatological regression toward the Apocalypse. Much of the course of modern history is the struggle of these two teleological narratives: one optimistic, progressive, and constructive, the other pessimistic, regressive, and destructive. Whether time bears out a positive or negative result for the future of humanity is at the root of the debate.

8 *Introduction*

Similarly, the study of ontology informs our perceptions and organization of space. Broadly speaking, ontology is the study of being, and the relation of things to other things. Inherent in the study of existential relations is the question of hierarchy. In physics, ontological hierarchy is a matter of scale which relates the subatomic particles to the universal cosmos. In biology, ontological hierarchy relates single-celled organisms to the biosphere. In computer science, ontological hierarchy relates computer codes, prompts, and commands to programs, which integrate entire systems. In geography, ontological hierarchy relates particular places to an overarching power structure, which monitors and regulates them in a systematic fashion. It does so through mapping, which is a form of surveillance bound up with the appropriation of knowledge for the purpose of authoritative control. Places are mapped for the purpose of hierarchical regulation and ordering.

This systematic ordering, policing, and regulation of space has become so pervasive with the onset and development of modernity that the progressive model of teleological time has become increasingly difficult to implement and realize. The modern global capitalist model is inherently degenerative and exhaustive with regard to what its own organizational language refers to as both human and natural resources. This capitalist market is enforced by the nation-state, and reinforced through technological pervasiveness. As a result, the triangulation of control of modern geographical space is designed to prevent any constructive temporal change toward progressive teleology from occurring. The result is a feedback loop of superficial change, while the processes that maintain it move toward an increasingly degenerative conclusion.

That feedback loop of superficial change is enforced "from above," primarily as a means to scramble the clarion calls for substantial changes to political, economic, and social conditions to improve along progressive lines that come "from below." Change is a new product; choice is a range of them. Both are mere commodities, and reinforce the very system that perpetuates teleological stasis through the enforcement of ontological spatial hierarchy. The concept of "technocratic monoculture" is useful in illuminating the ubiquitous nature of this reality, along with its implications as an increasingly expansive and pervasive condition. Like modernity, which is arguably a western construct, it has a global effect, and its expansive reach is something that becomes increasingly difficult to challenge on its own terms, a new Leviathan that simultaneously traps its own participants in a cycle of production and consumption that benefits a few at the expense of the many.

The comprehensive nature of the technocratic monoculture's control of space is evident in the way it appropriates temporal events of the past that either embody or at least introduce teleological progress. Monuments and the historical narrative they embody are constructed in a way that serves to spatially reinforce the ontological hierarchy, even if the event in memoriam was intended as a progressive teleological challenge to it. Here, the control of space in which events happen becomes a proxy control of time, and the systemic nature of ontological hierarchy reaches across the time-space continuum to become even more complete. The nature of this control is rooted in imperialist practices, which seek to render space and the people who inhabit it passive, so that the commodification of the natural and human resources therein can be fully exploited. To demonstrate this tactic, I explore the implementation of the U.S. military's "Clear Hold Build" as the apotheosis of imperialistic ideals and design.

The scale and scope of the technocratic monoculture within the context of modernity is an overwhelming prospect, yet the drive for progressive teleology is as strong as ever. From the global democratic revolutions in Congo, Czechoslovakia, France, Mexico, Senegal, and the United States in 1968, to the 1989 uprisings in South Africa, China, and the Soviet Union and the Eastern Bloc, to Seattle in 1999, to the Green Revolution in Iran in 2009, to the Arab Spring in 2010, and the global Occupy Movement of 2011, there is a growing sense that the present "order of things" is both undesirable and unsustainable. Overt revolutions, however, are an increasingly dangerous and untenable proposition. I suggest that political, economic, and cultural conditions of "unmappable places," which are impossible for the ontological hierarchy to triangulate, bind, and control, are the best staging grounds to build the foundation for progressive teleological change to occur. Neighborhoods, communities, and regions are places where unregulated political, economic, and social conditions can manifest, "from the bottom up." Because these places are defined from within, not from without, they are difficult for the agents of technocratic monoculture to identify, regulate, and control. Rather than direct confrontation with the state and the military-industrial complex at its disposal, simple disengagement becomes a viable and entirely legal alternative, given its commitment to choice as a primary market imperative. Here, teleological progress can occur within the same sociospatial matrix of ontological stasis as technocratic monoculture. Unmappable places allow for the space necessary for the temporal progression of history along teleological lines to occur. These are the places in which we can redefine value, worth, power, progress, and change. There is time, and now there is space for an alternative to the present course to manifest.

As such, ours is to consider the implications of the systematic enforcement of spatial order which fractures the possibility of a historical teleology that connects Hesiod to Hegel, Marx, and Engels: namely, that our common humanity is undermined by exploitative, hierarchical systems which prevent us from realizing the spirit of shared human interests, and to consider how humanity could ultimately undo these systems and return to a "Golden Age" of mutuality and commonwealth. To do so, ours is to first examine the epochal context of our present situation, namely the conditions of modernity as an acceleration of this fundamental conflict. Chapter One will consider the question of modernity as a historical concept. Chapter Two explores how the elements of modernity affect our perceptions of time and space. Chapter Three examines the globalization of institutional hierarchy within the context of modernity, and considers the resulting phenomenon of technocratic monoculture and its consequences. Next, towards a consideration of the implications of hierarchically ordered space and its inherently repressive functions, Chapter Four examines the role of historical events in shaping the course of history and how their spatial appropriation by the technocratic monoculture serves to halt temporal progress in its tracks. Finally, in the conclusion, is an assessment of the current state of the ongoing Titanomachy to determine how best to proceed as combatants in this metaphysical battle against Zeus and Olympian order.

FROM MYTHOS TO TECHNE

Time and space have both natural boundaries and humanly defined boundaries. Plato described the difference in parts of his work, which have been collectively understood as "the theory of forms." Here, matter has a definite corporeal essence, which is objective and substantial. Our constructions, perceptions, and rearrangements of matter, inherently subjective and ideological, are a product of our intellect and understanding, which Plato qualified as forms. Within the framework of modernity, the humanly defined boundaries of time and space, conceived as assertions of power and control over territory, including whatever and whoever occupies its space, have come to supersede the natural boundaries. Forms replace matter; *techne* replaces *mythos* as the defining factor of human relations.

Michel Foucault explained the comprehensive nature of the term *techne*. Arguing for the inclusion of spatial design as a factor which informs historical processes, he suggested: "I think that it should be much more along the lines of that general history of the *techne*, rather than the histories of either the exact sciences or the inexact ones." He followed that in applying "this general history of *techne*, in this wide sense of the word, one would have a more interesting guiding concept than by considering opposition between the exact sciences and the inexact ones."[13] Within this framework, a wide-ranging consideration of a subject as ominous and unwieldy as modernity can begin to come into focus.

The matters of time and space are obfuscated; the conceptual forms of time and space then become the reality in which we act and interact on a regular basis. In the process, the natural boundaries of both have been obfuscated for the purpose of control, exploitation, and appropriation. This process has become exponentially more comprehensive with the onset of modernity, which is ultimately the subject of this inquiry. At its base, this book is an attempt to understand how the condition of modernity affects our perceptions of time and space. Its primary interest is to understand the development of spatial triangulation for the purpose of order and control, which is intended to prevent anything unregulated, unmonitored, or unsanctioned by hierarchical authority from happening.

Throughout this consideration, I want to examine the effect that modernity has had on the ways our globalizing world conceives of, processes, and understands the fundamental concepts of time and space. While modernity may be a relatively new phenomenon in the history of the world, one that is not even fully understood by its own theorists, time and space have long been understood by humanity as elemental in our understandings of who we are and what we value. Together, they form the continuum that Isaac Newton declared the metaphysical foundation for classical mechanics, and by proxy, the very basis of the human condition. Both time and space are inherently battlegrounds on which the struggle for liberation, freedom, and fulfillment occur on a fundamental level. To understand the implications of these conditions, we have to ask some very basic questions: What is time? What is space? What is modernity? How has modernity affected the conditions in which time and space occur along with our perception of them? Have the social, political, and economic forces that operate within the context of modernity allowed for human liberation in temporal and spatial contexts, or do those forces functioning within the framework of, and shaping the parameters of, modernity make for inherently oppressive conditions within it?

What is the relationship between historical time and geographical space in the context of modernity? The question is not so much a novel one as it is a basic one. While a general approach to the question would produce a superficial answer, a comprehensive approach is impossible. As such, answering the question becomes a matter of identifying particular intersections of time and space by considering history from a thematic perspective, and separating the constants from the variables to identify what conditions the developments in question have in common. Ultimately, I argue that the triangulation and regulation of geographical space often challenges in order to prevent the dynamics of historical change from materializing. Historical change often manifests as theory first; it is only in the realization of ideas in physical space that theories become practice. With the onset of modernity, the dynamic of temporal change becomes increasingly difficult to realize. There is less "open" space for new social theory to manifest, therefore, less opportunity for historical change to occur. Modern forms of spatial triangulation that promote the hierarchically imposed order of technocratic monoculture prevent unregulated temporal change, which promises liberation from its more oppressive qualities. In the meantime, the processes of modernity enforce temporal and spatial stases. Teleology is interrupted; modernity prevents substantive change, which would undo the impositions of degenerative hierarchical forms and reclaim the essential nature of both time and space from occurring. Destructive political, economic, and cultural practices hold sway over constructive alternatives.

NOTES

1. Hesiod, *Theogony/Works and Days,* trans. M. L. West, (Oxford: Oxford University Press, 1988), 3.
2. Charles Seeger, "On Dissonant Counterpoint." *Modern Music,* Vol. 7, No. 4 (June–July 1930): 25–26. Seeger argued that "the effect of this discipline" was "one of purification."
3. Hesiod, *Theogony/Works and Days,* trans. M. L. West (Oxford: Oxford University Press, 1988), 40.
4. Robert Graves, *The Greek Myths: Vol. 1* (New York: Penguin Books, 1955), 40–41.
5. Hesiod, *Theogony/Works and Days,* trans. M. L. West (Oxford: Oxford University Press, 1988), 40.
6. Pausanias, *Description of Greece with an English Translation,* trans. W.H.S. Jones and H. A. Ormerod (London: William Heinemann Ltd., 1918), ix. 25.4.
7. "CCTV Cameras (CYCLOPS)," *Power Telecomm, Inc.,* Accessed May 12, 2014. http://www.powertelecomm.com/product/cctv.html
8. Helen Morales, *Classical Mythology, A Very Short Introduction* (Oxford: Oxford University Press, 2007), ii.
9. Ferdinand de Saussere, *Course in General Linguistics* (New York: Columbia University Press, 2011), 16.
10. Charles Sanders Pierce, *The New Elements of Mathematics,* ed. Carolyn Eisele (Atlantic Highlands: Humanities Press, 1976), Vol. 4, 54.
11. Ferdinand de Saussere, *Course in General Linguistics* (New York: Philosophy Library, 1959), 67.
12. Georg Wilhelm Friedrich Hegel, *Lectures on the Philosophy of History* (London: G. Bell and Sons, 1910), 238.
13. Paul Rainbow, "Space Knowledge and Power," in *The Foucault Reader,* ed. Paul Rainbow (New York: Pantheon Books, 1984), 256.

1   Modernity and Its Discontents

INTRODUCTION

Infinity and eternity are inevitably bound by human comprehension. The word "infinity" is derived from the Latin *infinitas*, which means "unboundedness," a concept influenced by the Greek concept of *apeiros*, or "endless." The idea is most regularly applied to the study of physics when conceptualizing cosmology, and hence has inherently spatial properties. Eternity is a temporal indicator which embodies past, present, and future. To comprehend space and time, philosophers, theologians, historians, and geologists subdivided both, ordering and rationalizing them in the process. Hesiod's "Ages of Man" were meant to bring a sense of temporal continuity to the relationship between the past, present and future. *Works and Days*, another of Hesiod's poems, lists five ages in succession: Golden, Silver, Bronze, Heroic, and Iron.

In Hesiod's model, The Golden Age was the first, symbolic of an ideal past. During the time of Kronos, Hesiod describes a sort of primitive communism, where there was no suffering and plenty of life's necessities to go around. Humans lived in harmony with each other, the environment, and the gods. The Silver Age followed the overthrow of Kronos by Zeus, in which men were reduced to prolonged adolescence. Due to their dependence on their mothers, once they left home "they lived but little time, and in suffering, due to their witlessness." Next was the Bronze Age, in which "a terrible and fierce race . . . with acts of violence . . . were laid low by their own hands." Hesiod concludes that "dark death got them, and they left the bright sunlight." The Heroic Age followed the Bronze, during which time "ugly war and fearful fighting destroyed them." Hesiod believed that the people of this age prior to his own fought the Trojan Wars, when "some of them were engulfed by the consummation of death, but to some Zeus the Father, son of Kronos, granted a life and home apart from men, and settled them at the ends of the earth." There, on the periphery, these demigods "dwell with carefree heart," and were regarded as "fortunate heroes." Lastly, Hesiod described his own age, The Iron Age, by wishing he had not been born in it. During this age, "men . . . will never cease from toil and misery by day or night, in constant distress, and the gods will give them harsh troubles." During this time, social order breaks down: "there will be no thanks for the man who abides by his oath or for the righteous or worthy man, but instead they will honor the miscreant and the criminal," and "there will be no help against evil."[1]

Hesiod's narrative is one of temporal degeneracy; things were initially good and got worse over time. This is a familiar round; "the good old days" is a colloquial expression often used to engage an idealized past held up against a deteriorated present. Other forms of religious discourse suggest that this decline is part of a cycle which would eventually bring regeneration and renewal. The Jainist cosmogenic cycle conceives of time as a twelve-spoke wheel: each spoke is an age. The first set of six *avasarpini*, not unlike Hesiod's narrative, is descending. During this time, the condition of the world deteriorates and misery is proportional. According to this temporal conception, ours is the fifth of the descending series, Kali-Yuga, which began in 522 BC. The mythologist Joseph Campbell wrote that "it is a period of unmitigated and gradually intensifying evil." Where the Jainist narrative breaks from Hesiod is in its conception of an eventual ascension from these depths. The upswing in the wheel will inevitably bring about another downturn, and the cycle will continue in perpetuity.[2]

In Christianity, the past as ideal is both temporally and spatially designated. The first verse of the first chapter of the Book of Genesis opens "in the beginning," (*Alpha*); God's first temporal action is to create terrestrial and extraterrestrial space in the forms of Heaven and Earth. Humankind is eventually cast out from the Garden of Eden for failing to abide by the rules laid out by its creator, an event commonly referred to as "the fall of man." Thus begins the pattern of temporal degeneracy so common in mythological conceptions of past and present. An ideal once existed, and has since been lost. God then sends his only son, Jesus Christ, to redeem humanity through his own sacrifice, which prepares the way for "final judgment." It would not be until the divine judgment of the living and the dead, the impending apocalypse which Saint John outlined in the Book of Revelation, and the destruction of the degraded world (*Omega*) that the opportunity for renewal could come. The renewal would not be material, however, as John the Revelator describes in the tenth verse of the twenty-first chapter: "And he carried me away in the spirit to a great and high mountain, and showed me that great city, the holy Jerusalem, descending out of heaven from God." As such, the ideal lost in Eden is only attainable by death, in nonmaterial form. In the Christian worldview, there is no hope for the living. The best humanity can do is hasten the apocalypse and prepare for final judgment (*Omega*). There is no possibility of return to a "Golden Age."

The Venerable Bede expanded on this narrative in his eighth-century treatise *The Reckoning of Time*. Arguing that the point of temporal origin of the universe is essentially the unknowable essence of God, he articulated six ages of the world based on biblical reckoning. Like Hesiod, Bede situated the present condition of humanity in the most debased age. Unlike Hesiod, he explicitly articulated that this degenerated age would have eschatological implications. Bede explained that "The Sixth age, which we are now in, is not fixed according to any generations or times, but like senility, this age will come to the death of the whole world."[3] By Bede's account of time, the Seventh and Eighth ages are postmortem, on another plane of existence altogether. He identified the chaos of earthly temporal life as a fundamental difference from the order of eternal afterlife: "And so our little book concerning the fleeting and wave-tossed course of time comes to a fitting end in eternal stability and stable eternity."[4] Here again, the only hope for humankind in the linear model of Christian eschatology is eternal salvation in heaven; there is no hope on Earth.

Regardless of the narrative form it takes, periodization appears throughout the history of temporal discourse, rooted in religion and mythology. Ages, epochs, eras, and periods help to comprehend and frame the past in light of the present in a rationalized, ordered way. At the same time, they arguably reduce the complexity of the time in question and obliterate any sense of continuity between the past and present. Often, they tend to romanticize the past as an ideal that has been lost and forgotten over the course of time. The concept of "the golden age" is a colloquialism that resonates from considerations of piracy, automobiles, film, and a range of subjects of popular interest. Its usual reference of a bygone era suggests that the qualities that made the ideal possible have somehow been lost. That the ideological roots of an ideal past and a debased present are in the foundational myths of many major religions shows what a powerful hold the discourse has on the human mind, and begins to explain how ideas of human degeneracy have shaped the course of history and our perceptions of it.

ORIGINS OF THE TERM

As theologians have constructed understandings of "The Ages of Man" to elucidate religious doctrine, so too have historians conceived of historical ages to explain the arc and trajectory of past circumstances. The periodization of history has changed and evolved over time to reflect the course of human events, as well as the attitudes and interests of historians. History tells us as much about the time in which it was written as it does about the past subject matter it considers. In this context, the subject of historical periodization invites critical analysis as it lays a rudimentary framework for more exacting considerations of particular times and places therein.

Any understanding of modernity as an historical epoch is bound up with larger questions of culture and subjectivity. Much of the formative discourse that sprung up around the concept of modernity was conceived in Europe from the fifteenth through the eighteenth century, and as such is subject to the beliefs and biases of the culture in which it was imagined. Modernity was largely attached to an ascendant view of western civilization as something distinct from and superior to the rest of world history. The exclusive nature of "Western History," loosely constructed in the debates over curriculum as ancient Greece and Rome, medieval Europe, and modern Europe and America, is a quality that exposes the subjective limits of the inquiry to conform to a predetermined narrative of hegemonic superiority.[5] The question of modernity's origins is a subjective one, but the conditions of modernity in the context of globalization or time-space compression are something that the academic community has a certain obligation to understand, if for no other reason than its comprehensive proliferation.

In this context, the most basic bifurcation of historical discourse occurs, ironically, at the geographical level. One camp of historians works within the tradition of western civilization. Historians working within this tradition collect their material from the history of Greece, Rome, medieval and modern Europe, and the remnants of its imperial expansion from the fifteenth through the nineteenth century. These times and places provide the content for historical analyses largely preoccupied with the impact that western civilization has had on the world. Up through the late twentieth century, these historians built a largely self-congratulatory narrative on the superiority of western cultural, political, and economic development. Here, history serves as inspiration. Myriad titles on the subject of civilization, all with very similar tables of contents and indexes, were published throughout the late nineteenth and early-to-mid-twentieth centuries. These are the stories of emperors, kings, presidents, and generals predicated on the idea that great men make history and the rest of humanity are passive subjects to the conditions they impose.

The other camp incorporates western civilization into a larger discourse of world history. This discourse emerged from the liberalization of the academic profession, a result of the post–World War II G.I. Bill, which gave working class citizens an opportunity to pursue academic degrees, and the civil rights movement, which saw an increased number of ethnically and racially diverse men and women enrolling in universities that had previously excluded them from studies. World history challenges the discourse of western civilization by broadening the geography of the subject matter to make it more inclusive. Historians building a world history tradition deliver a challenge to the exclusivity of the western civilization tradition. World history includes the histories the western hemisphere, East Asia and sub-Saharan Africa, which are conspicuously absent in the western civilization tradition until Christian "explorers" arrive. The narrative of world history tends to be less congratulatory and more critical than the western civilization model in its consideration of the past. In this context, proponents of world history argue for the legitimacy of precolonial nonwestern cultures while incorporating the arrival of western civilization and the forces of modernity into the larger narrative. In doing so, they expose some of the more difficult paradoxes of western society. One of many complications that arise in this debate is western civilization's commitment to the political discourse of freedom, while having built much of its wealth through the institution of slavery. The debate has shaped our most basic understandings of how we view the past, and, by proxy, how we view ourselves in the present. School history standards have been informed by this debate since the 1980s, the nature of which historian Peter Stearns has described at various points in his analysis of it as belligerent, intolerant, unyielding, antagonistic, and troubling, and ultimately as a conflict having gone too far.[6]

As it stands, historians working within the context of the western tradition inherit a construction of the past divided into three distinct epochs: ancient, medieval, and modern.[7] From these three main divisions, myriad subdivisions exist. The history of the ancient world can be divided by the temporal geography of power: first Greece, then Rome. Or, it can be divided by the development of technology: Stone Age, Bronze Age, and Iron Age. The medieval period is particularized into the early, middle, late, and high periods, and modernity has its own early and high phases, only to be complicated further by the discourse surrounding the theoretical construct of postmodernity. The increasing refinement of periodization has become in and of itself a phenomenon that exponentially complicates the understanding of history. Whether or not these categories contribute to or obfuscate our understanding of the past is as much a matter of debate as the nature of the categorization itself.

Proponents of world history often superimpose this model on their own subjects, building a thematic case for the continuity of historical conditions that resonate outside of the western tradition. Hence, China's *ancien regime* compares to that of Europe's, feudal Japan has similarities to feudal Europe, and the imperialist tendencies of sub-Saharan African kingdoms seem analogous to those of medieval Europe. This thematic continuity becomes more complete with considerations of modernity as the comprehensive nature of the modern age (or at least our academic understanding of it) has an unprecedented conditional unity to it that tends to subsume any contest about historical ascendancy. Regardless of its origins, modernity as a pervasive cultural, political, and economic condition is a resonant element of the human condition, and to either be celebrated or challenged, it first has to be understood.

Modernity is a familiar word, but an unfamiliar concept. Its Latin root, *modus*, means "now," "now existing," or "pertaining to the present." The Roman politician Magnus Aurelius Cassiodorus was the first to use the Latinate predecessor of modernity, *moderni*, in AD 510. The term appears in a letter to the Roman Consul Importunus, in which Cassiodorus describes the young politician's father and uncle as "men who adorned the modern age with antique virtues."[8] Here, Cassiodorus juxtaposed the past and the present, painting the former as ideal, and something worth preserving when compared to the latter. The past, *antiqui*, was something to be recovered and applied to the present. In the context of Cassiodorus's time, with the rise of the Gothic kingdoms changing the historical dynamic of the Roman Empire, modernity represents change. As such, Cassiodorus, like Hesiod over a thousand years before him, offered an idealized view of the past as something preferable to the chaotic swirl of the present.

In historical discourse, the idea of "modernity" as an era distinct from the past is largely bound up with the intellectual shift from Scholasticism to Humanism in fourteenth-century Europe. Scholasticism was the preeminent form of scholarly inquiry in Europe since the development of the university in the twelfth century, and was largely a discourse preoccupied with defending Christian beliefs within the context of academic inquiry. Scholars such as Petrarch and Giovanni Boccaccio began to mark an ideological turn from the comprehensive nature of Christian dogma beginning in the fourteenth century, resulting in a shift in academic approach that came to be known retrospectively as "Humanism." Where Scholasticism was preoccupied with theology as a means of explaining the past, Humanists focused on the agency of human beings; Humanism's focus on empirical evidence instead of faith marked a shift in ideological approach to knowledge and understanding.

While historians working within the context of scholastic thought had written about the past with the outcome of degeneracy and apocalypse a foregone conclusion, Humanist historians separated themselves from that line of thought and suggested a new age, reframing the past in light of new understandings of antiquity and the possibility of a reinvention of the human condition, or *renaissance*. The Italian scholar Petrarch began this temporal and ideological reconfiguration by suggesting in his 1348 *De Viris Illustribus* that the past had three, distinct periods: the idealized age of Roman empire, a period following the fall of Rome, which he problematized as a "Dark Age," and a renovation and reconstitution of the "Golden Age" which he believed humanity was just beginning to experience.[9] In 1442, Leonardo Bruni elaborated on this tripartite framing of history in which the fall of Rome marks the end of a primary historical period (for Bruni, this was in AD 476 with the deposition of Augustulus), a second interstitial period between the fall of Rome and the establishment of a new order which marks the third period beginning in the late thirteenth century.[10] Bruni's three-part framework of history as ancient, medieval, and modern was formalized over time and became institutional in the eighteenth century with the publication of Christoph Cellarius's *Universal History Divided into Ancient, Medieval and New Period*.[11]

WHAT IS MODERNITY?

If modernity is something distinct from the medieval and ancient epochs, what then are the characteristics that define the modern age? The word "modern" evokes stylized visions of stainless steel appliances in gleaming homes, effortlessness, and hyperdesign. But what is modern: the latest computer design, or electricity itself? Are the systems that enable the inception and refinement of design the basis for modernity, or are the refinements themselves the basis for our conceptual understanding? In the 1950s and 1960s, arguably the peak of what we can consider the "Modern Age," and certainly the default image that most people assume when they are asked to envision "modernity" as a phenomenon, the discourse of what was modern centered on the development of laborsaving machines, devices, and gadgetry. While this represents a popular understanding of what constitutes modernity, it is far from comprehensive.

As the question of modernity as an epoch is arguably a relative one, this consideration focuses on the development of particular economic, political, and social forces from roughly the fifteenth century onward that mark the parameters of the evolutionary shift that Petrarch first envisioned in the fourteenth century. Consider four themes as the fundamental elements of modernity: *technology, urbanization and nationalism* (dual forces which, taken together, represent fundamental changes in the organization of human populations), *the proliferation of market economies*, and *the presence of a democratic ideology*. While these four institutions have had a global presence in some form or another since well before the fifteenth century, it is the exponential proliferation and advance of their conditions that arguably marks the dawn of the modern age.

Technology, as previously mentioned, is the most elemental and accessible touchstone of modernity. The various forms of technological development since the fifteenth century, and their exponential expansion, provide the foundation for the other three themes to geographically proliferate over the course of modern history. The evolution of communication technology, transportation technology, military technology, and information technology are foundational to the form and function of the modern world. From the laying of the first transatlantic telegraph cable in 1858, to the omnipresence of digital information over global, wireless networks in the twenty-first century, the spread of communication technology has been instrumental in the shaping of the modern world. The incremental development of transportation technology, from regional patchworks of canals and railways, to the near singular unity of transportation networks on a global scale, has also shaped contemporary perspectives of modernity over that same period. The evolution of military technology, from cavalry charges at the beginning of World War I to the decisive power of the atomic bomb which ended World War II, has had an enormous impact on the condition of modernity. Information technology, from the mid-eighteenth-century attempt by Denis Diderot to systematically catalog human understanding in *Encyclopedie*, and Carolus Linneaus's taxonomical accounting of the natural world in his 1735 work *Systema Naturae*, to the categorical immensity of knowledge available through the internet, the development of information technology has also had a profound impact on the social, political, and economic conditions of the modern age.

The dual forces of *urbanization and nationalism* together have played a large role in shaping the conditions of the modern era. While the history of cities like Athens, Carthage, Rome, and Alexandria define much of our understanding of the ancient world, after the fall of Rome, urban populations fell drastically in the European world. Historical and archeological evidence of cities thriving outside of Europe during its medieval period exists in the Americas at Tenochtitlan and Kahokia, in Africa at Great Zimbabwe and Kiguba, and Asia at Xi'an and Tamralipta, but even with this, for the course of history, most of humanity has lived in rural conditions. It was not until 2009 that the majority of the Earth's human population lived in cities.[12] Urbanization has become a defining feature of modernity, yet these cities all operate within the context of national identities. Simultaneously, nationalism, a system of spatial organization formed of early nineteenth-century counterrevolutionary conservatism, drew exacting political boundaries, often at the expense of preexisting cultural and economic conditions, in an attempt to fundamentally redefine the identities of person and of place. Nationalism is the principle basis of international relations today. Nations are sovereign political entities with the power to make war, regulate trade, and pass laws.

The proliferation of market economies is a fundamental characteristic which makes modernity an epoch distinct from its temporal predecessors as well. To be sure, market economies have existed in one form or another since prehistoric times, and the history of economic trade is rooted in the most basic concepts of supply and demand. Still, the proliferation of market economies in the context of globalization is a phenomenon that is inherently modern and distinct from the past. This is not to say that globalization is a necessarily new phenomenon. We see its roots in the Babylonian *tamkarum*, the North African Sahel, the thirteenth-century travels of Marco Polo, the expedition of Zheng He in the early fifteenth century, and the voyages of Christopher Columbus which punctuated the expansion of global trade routes at the end of the fifteenth century. Whereas markets have always been a driving force behind cross-cultural exchanges, the emerging presence of the market as a monoculture is one that is not without implication, as global capitalism expands at exponential rates in pursuit of profit.

The force that has historically served to keep the previous three conditions in check is *the presence of a democratic ideology*. Here again, democracy is not a modern invention. Evidence of democratic ideology is present in the Athenian Agora, the Haudenosaunee Longhouse, and the Icelandic Althingi. The global proliferation of democratic ideology is the thing that makes modernity distinct from other ages. This is not to say in practice, but in theory, democratic ideology has been a vital element of modernity, and continues to be today. It is the philosophical fuel of revolutions, from Putney to Port Au Prince in the seventeenth and eighteenth centuries through Tiananmen and Tehrir Square in the twentieth and twenty-first centuries. Democratic ideology provides the conscience, the moral foundation, the ethical substance that holds technology accountable, which keeps nationalism from manifesting as systematic injustice, and which insists that market economies operate fairly and openly to provide opportunity for all rather than limiting it for the exclusive benefit of a few.

These four forces, taken together, have had a profound impact on the course of history since the acceleration of their processes in the fifteenth and sixteenth centuries. Whether we view history from a western or global view, the conditions of modernity continue to expand. Their influence on the world, inside of the western tradition and outside it, is evident. Technology is increasingly expansive, cities continue to grow, national borderlines cross over the planet, and protests around the world ring with revolutionary calls for democracy. As the origins of these phenomena within the historical frame of "Western Civilization" are debatable, their exponential global proliferation is not. Places on the planet that have not been affected by the conditions of modernity in one way or another are increasingly uncommon.

COMPLICATING THE CONCEPT

It is worth reminding ourselves of the subjectivity of historical discourse at this point. As E. H. Carr warned: "History means interpretation."[13] We are arguably living in "The Modern Era" or "The Age of Modernity," and the immediacy of the experience shapes and colors our perception of it. Writing a history of a present age is different than writing the history of a past period or course of events that resonates in the present. The temporal distance between past and present affects the resonant quality of the subject. Classical historians are constantly apologizing for a lack of sources, and qualifying those which exist as partial, biased, and incomplete. Modern historians have an opposite problem; the immediacy of the subject results in an overwhelming amount of source material. What to use, what to leave out, and what to make of it all is inherently subjective. As such, we do well to be mindful of our temporal immediacy with regard to such matters. Historical considerations of modernity are in some ways histories of the present, even if they are examining matters a few hundred years old. The geographer Yi-Fu Tuan suggests as much in observing the paradox of distance in *Space and Place: An Experiential Perspective*, writing that "thought creates distance and destroys the immediacy of direct experience, yet it is by thoughtful reflection that the elusive moments of the past draw near to us in present reality and gain a measure of permanence."[14]

Much of the way we conceive of modernity comes out of the discourse of the Enlightenment, the self-congratulatory title bestowed upon itself by the eighteenth-century intellectual movement which, building on Humanist and Renaissance developments in the construction of knowledge and academic inquiry, framed the world in terms of the lightness of knowledge and the darkness of ignorance. The foremost historian of the period, Peter Gay, calls the Enlightenment a "Prelude to Modernity," in which "fear of change was giving way to fear of stagnation; the word *innovation*, traditionally an effective term of abuse, became a word of praise."[15] As we have seen by considering the formative works of Petrarch, Bruni, and Cellarius, modernity began as an inherently an intellectual construction of the present within the context of a historical framework. Modernity can only be understood in comparison to something that is premodern; that is to say that the characteristics of modernity depend on a preceding condition with which to contrast it. The ideas of change and innovation, which Gay suggests are elemental to the discourse of modernity in its formative period, are tools with which the modern present distinguishes itself from the premodern past.

The counterpoint to arguments about exponential temporal change in the context of modernity, that more things are happening at a faster pace, is that the type of change modernity theorists are preoccupied with is merely superficial, and stasis is the overarching condition of modernity. Albert Einstein argued in his theory of relativity that time dilation is a physical process in which the increased speed of motion actually slows the passage of time. Michael Lind's neologism "turboparalysis" encapsulates the idea that calls for political, social, and economic change result in merely superficial affectations: "years of activity without action and motion without movement."[16] It is easy to argue that modernity has accelerated the processes of spatial reckoning toward a point of temporal and phenomenological stasis. The result is a feedback loop of superficial change without a substantial shift in the processes which drive it.

Some scholars question whether modernity is even a concept worth articulating in the academic sense. Jack Goody calls into question the periodization of history into ancient/medieval/modern, suggesting that these are western inventions superimposed onto world history "as part of a comprehensive process" rather than viewing world history in a more inclusive and complex way.[17] In his 2006 work *The Theft of History*, Goody asserts that the exclusive nature of modernity as a concept is inherently flawed: "'Modernity' is conceived as a purely western phase, but even the criteria for its emergence, though stated in categorical terms, are far from clear."[18] Peter Van Der Veer acknowledges the overwhelming force of western modernity, but argues for a plurality of modernities as each culture experiences its social, political, and economic conditions with its own history on its own terms.[19] Both of these arguments occur within the context of the western/world history debate from which any consideration of modernity springs.

To further complicate the question, C. A. Bayly suggests in *The Birth of the Modern World 1760–1914* that modernity is both a process and a period. The process of modernity involves a significant degree of social understanding; according to Bayly, the modern world became modern in the nineteenth century as a result of events emanating from the western world (broadly speaking: the transatlantic European community) "because a considerable number of thinkers, statesmen, and scientists who dominated the ordering of society believed it to be so."[20] Perception of modernity is a significant component of its epochal status; it is an inherently subjective condition. The degree to which something is modern or not is a matter of judgment, based on criteria constructed and designed by hierarchical agents of the system such inquiry sets out to examine in pursuit of a particular rhetorical interest.

Is modernity something distinct from the ancient and medieval epochs of history, or is that periodization altogether problematic? Is our present hindered by the past, or are we headlong toward the future? If the former, what are the hindrances; if the latter, what does the future look like? If nothing else, it is the speed at which these institutions have advanced since the fifteenth century which arguably what makes modernity a distinct historical epoch. Indeed, speed is a fundamental characteristic of modernity. To that point, the historian William Everdell writes that "there is good reason to wonder if 'modernity' means anything at all beyond a change in the pace of change."[21] Jose Moya calls the phenomenon of accelerating human reproduction, production, transportation, financial flows, consumption, and aspirations "massification," and argues that it forged the modern world.[22] These historical theories echo the technological theory of Gordon E. Moore, who proposed Moore's Law in 1965, which asserts that computing technology reproduces itself exponentially by doubling every two years.[23] The futurist Ray Kurzweil expands on this idea in his 2001 essay *The Law of Accelerating Returns*, stating that "technological change is exponential, contrary to the common sense 'intuitive linear' view."[24]

Most scholars generally agree, if for no other reason than for the basis of a debate, that modernity becomes a defining conceptual factor in understanding the human condition beginning in the fifteenth century. To Everdell's point regarding the temporal framework of modernity as a historical construction, another circumstance scholars generally identify as unique to the modern experience is the acceleration of these aforementioned primary indicators. In this context, acceleration and massification underlie the temporal and spatial framework of modernity in a fairly comprehensive manner. Anthony Giddens writes, "the modern world is a 'runaway world': not only is the pace of social change much faster than in any prior system, so also is its scope and . . . profoundness."[25]

This pace of change, exponential rather than linear for the first time in human history, is in and of itself a phenomenon worthy of attention. In 1970, Alvin Toffler began to explore the implications of exponential change in his book *Future Shock*. He argued that the pace of change in technological modernity was overwhelming and disorienting. As a result, this pace of change had pushed the modern world out of the industrial and into the "super-industrial era."[26] The conditions of the new era, as Toffler predicted, were indicative of a comprehensive change. The institutionalization of disposability and obsolescence, both with products in the home and with people in the workforce, would result in a deep sense of alienation that would require new ways of thinking about everything from family structure to food distribution.

Since 1970, when Alvin Toffler released his book *Future Shock*, the pace of change has only quickened, prompting the 2013 follow up by Douglas Rushkoff: *Present Shock: When Everything Happens Now*. Rushkoff's analysis focuses on the omnipresence of media, which makes information overwhelming and lacking in any acute resolution. According to Rushkoff, the future is now, but we are ill-prepared to handle it in any comprehensive way. In our rush to gather information about the world around us at record speed, we lack the ability to process or make sense of it. The result is a lack of context, a lack of focus, and ultimately a lack of understanding about our present condition. This incoherence leads to an omnipresent sense of urgency about everything, and brings forth a longing for the only form of resolution that can come from what we perceive as absolute chaos: apocalypse.[27]

It is in this framework that atmospheric chemist Paul Crutzen takes the concept of temporal periodization a step further, suggesting a new geological epoch as altogether distinct from the present construction of the Holocene. The Anthropocene Epoch, according to Crutzen, takes into account human influence on the earth and its environment. To change the geological chronology to account for this, the transformation must be substantive, and there must be a pedagogic purpose to it. The root of the word Holocene is *holo*, which is Greek for "whole." Crutzen's proposal of departure from the Holocene is directly related to the concept of modernity, namely that the impact of human technology since the Industrial Revolution has had a detrimental effect on the biosphere.[28] In acknowledging the cumulative effect modernity has had on the human condition, the debate between scholars regarding the validity of the term and the usefulness of temporal periodization in the context of the western/world history debate seems entirely academic. The relationship is symbiotic. Regardless of its origins, the condition of modernity has a comprehensive effect on the world, and the world impacts the course and trajectory of modernity.

MODERNITY AND METATHEORY

Applying the prefix of "meta-" to an academic inquiry or invoking the concept as discourse invites a kind of wariness worth both acknowledging and addressing. The existential nature of the concept tends to invoke a certain abstraction that might initially seem impractical and indulgent, and therefore eccentric. Metatheory is essentially a theory of theory. To apply the concept of metatheory to both history and geography, the academic discourses of time and space respectively, is to suggest an essential theory of historical and geographical theory. Put more plainly, considering metatheory is to examine the self-referential nature of a discipline. Books about history and geography written in the nineteenth century tell us as much about the time they were written in as they do about the subjects they overtly consider. Metatheory unpacks this, and examines the context of ideas and arguments to better understand them.

Meta as discourse was first manifest in the subject of philosophy. Aristotle's work *The Metaphysics* is a treatise on existence and being. The title of the work is interesting enough; a work on philosophy that invokes physics. Meta in this case signifies something of a higher or second order. Aristotle's title suggests something beyond the physical order of things. *The Metaphysics* is essentially a consideration that aims to rectify the constancy of universal order with the perpetual change that occurs at all levels of existence: order and chaos. His argument was that there are certain constants of form among all the variables of matter that manifest over time and space. As these matters apply to philosophical discourse, Aristotle accounts for both when and where in his work on logic *The Categories*. Whether they attach to things in space or events in time, both are relational positions. Whereas theory is the consideration of these relational positions through time (history) or in space (geography), metatheory is a consideration of the elemental foundation upon which these movements occur.

The theological philosophers of the Middle Ages used Aristotelian metaphysics as a way of legitimizing the academic and religious discourse of Scholasticism, attributing all universal constancy to God. John Duns Scotus and William of Ockham turned classical philosophy into a theological exercise, merging the existential observations of Socrates, Plato, and Aristotle with the doctrine of the Roman Catholic Church. Philosophical discourse as metaphysics reemerged as an intellectual exercise in the Enlightenment, when Immanuel Kant reframed it by applying skepticism to the discourse. Kant argued that all physics take place in space and time, and that our knowledge of the events that occur there are "synthetic apriori," meaning that we construct our understandings of the physical world as such. These are not knowable things by Kant's estimation, and our estimation of what is beyond the physical manifestations of time and space is therefore little more than fabrication. Kant argued in *Prolegomena to Any Future Metaphysics* that "the task is difficult and requires a resolute reader to penetrate by degrees into a system based on no data except reason itself, and which therefore seeks, without resting upon any fact, to unfold knowledge from its original germs."[29]

Shailer Mathews wrote in *The Spiritual Interpretation of History* that "when a historian enters into metaphysics, he has gone to a far country from whose bourne he will never return a historian."[30] Mathews was writing about what were in his opinion the limits of traditional academic historical inquiry, suggesting that the spiritual element that the discipline lacked was both vital and personal. Without it, historians tended to dogmatize history, offering what they believed to be *the* interpretation of history rather than *an* interpretation of it. By incorporating the spiritual element, Mathews suggested that a historian may not only begin to understand the past, but also realize the "perspective of process" that exposes "direction and tendency," ultimately bringing to light "at least the general direction of the future."[31]

As such, it seems worthwhile to heed Mathews's warning and still move forward in applying a metatheoretical inquiry to both history and geography in hopes of exposing those connections between the past and present, and ultimately of the future. Much of this groundwork has been laid by scholars already; ours is to summarize their findings and apply them to the broader question of how systemic forces shape time and space, and how resistance forms and reacts to those same conditions. How does fundamentally questioning the conditions in which modernity happens reframe our own understandings of our current condition, and how do the answers offer an opportunity to change the things we cannot abide?

The concept of metahistory is one that the modern historian Hayden White has explored at length. In *Metahistory: The Historical Imagination in Nineteenth Century Europe*, White argued that historians do not just recount history, they make it by framing their subject matter in a narrative discourse. In doing so, historians impart ideology through an argument which they use to frame the subject of their work.[32] History, as a subjective discourse, is variable. According to White, history changes because historical perspective changes. The "modes of explanation" historians use to relate past events depend on the ideology of the writer and the intended purpose of the work.

Similarly, Martin W. Lewis and Karen E. Wigen have given thorough consideration to the subject of metageography. Building on White's concept of metahistory, Lewis and Wigen argue in their book *The Myth of Continents: A Critique of Metageography* that metageography is "the set of spatial structures through which people order their knowledge of the world."[33] They suggest that the discourse of academic geography produces an atrophied sense of space and place in a changing world. By deferring to geographical concepts which reinforce the hegemonic power and control of governments and corporations, nation-states and even continents, the discourse of geography serves to obfuscate rather than inform our understanding of the world around us.

Metahistory exposes the agendas of historians as agents of change, while metageography exposes the agendas of power structures resistant to change. Herein lies the thesis of this work, namely that the forces of historical development, which are dependent on certain degrees of agency which result in varying degrees of change that historians can evaluate and use as markers of progress against past circumstances, are triangulated and stifled by the forces of geographical ordering, which seeks to preserve and expand existing social, political, and economic structures for the purpose of reifying hierarchical power and control.

This is metatheory, a consideration of theoretical foundations, practical form and functional utility. The concept of metatheory as a discursive foundation for intellectual inquiry is often dismissed as a form of academic misanthropy. To be sure, *meta* as discourse can be a tedious exercise of linguistic trickery and doubletalk, but to dismiss its fundamental utility because of past practices is to err on the side of prejudice. Rather, engaging metatheory as a bridge to elemental questions about the nature of an academic discourse seems appropriate given the swirl of modernity in which no geographers or economists can agree on what "globalization" actually is, nor can historians seem to agree on what "modernity" is. Such naiveté begs for more functional forms of inquiry, even if they are initially engaged with a sense of chance and improvisation. To borrow from Duke Ellington's exposition on the early formation of his musical direction, "it's a good job we didn't know what we were doing."[34]

Rather than studying events that have manifested in geographical space over the course of historical time, consider instead the systematic development of space and time on their own terms. Theory studies the movements on a chessboard; metatheory considers the board itself, its inherent possibilities and its inevitable limitations. Chess, being an ancient and universal game, has value as a metaphor. As Benjamin Franklin observed in his 1786 essay *The Morals of Chess*, the game promotes foresight, circumspection, and caution; three good qualities for any academic pursuit and inquiry. Franklin also suggested that "chess is the habit of not being discouraged by present bad appearances in the state of our affairs; the habit of hoping for favourable change, and that of persevering in search of resources."[35] The applicability of metatheory is the applicability of chess. There is an inherent wisdom embedded inside of both that transcends the particularities of any one time and place. By considering the essential contexts in which those particularities occur, time and space, we can better appreciate the larger arcs and trajectories of human endeavor, and begin to either fix navigation or bear down on course as necessary.

ONTOLOGY AND TELEOLOGY

In *The Sophist*, Plato wrote down what the visitor from Elea had said to Theaetetus about the nature of ontological debate:

> Their dispute about being, in fact, has something of the effect of a war of giants against gods ... One party are for dragging down everything from the unseen heavens to the earth ... their antagonists in the dispute are so very cautious, and conduct their defence from some invisible height ... The battle over this issue between the two armies, Theaetetus, rages, as it has always raged, with unabated fury.[36]

Bringing the unseen down to earth, exposing the incomprehensible and making it accessible is the work of ontological inquiry. According to

Reinhardt Grossman, "Ontology asks and tries to answer two questions. What are the *categories* of the world? And what are the *laws* that govern these categories?"[37] Ontology thus considers the order of things, and the enforcement of that order. Put otherwise, ontology is a set of object or subject types and relations; it is a comprehensive system of classification and categorization. As the Eleatic described it, the enforcement of that order depends on a hierarchical power structure.

Ontological hierarchy is a term that ranges across several forms of disciplinary knowledge and understanding. Once categorized and classified, things (material or ideological), assume relations with other things. In the branch of philosophy known as metaphysics, ontological hierarchy is a hierarchy of existential relations. In the works of Aristotle and Plato, both the physical and ideological realms are ordered hierarchically, as matter and form. In mathematics, ontological hierarchy is used to reflect the complexity of interdependence in theoretical models. In biology, ontological hierarchy is inherently linked to evolutionary theory. In physics, ontological hierarchy provides a coherent description of the cosmos and a set of inferences which can be drawn from that description from universals to particulars. Comprehensively, ontological hierarchy categorizes and formalizes relations between objects and subjects within a particular system. That system, in terms of human relations, can manifest in political, economic, and cultural terms. The hierarchies that result from them are largely the product of power dynamics, upon which position in the ontological framework of any given set of conditions relies.

Applied to historical constructions of time, ontology exposes the systems in which societies are organized and ordered. Over the course of human history, the historical record accounts for societies which were and are fiercely hierarchical, and others less so. Societies that were or are more hierarchical concentrate power in political economic and cultural terms, whereas power is diffused in less hierarchical societies. Herein lays the difference between democracies and despotisms, egalitarianism and elitism, social justice and injustice, citizenship and subjection, and equality and exploitation. The geographical application of ontology shows the spatial effects of social organization and ordering. The history of imperial expansion is the history of hierarchical ordering of space along political, economic, and cultural lines.

The ontological condition Hesiod describes under the rule of Kronos is one in which time governs the reality of human existence. This "Golden Age" Hesiod describes is something humanity has since aspired to either return to or create anew. This regressive and progressive instinct exposes our cyclical and horizontally linear perceptions of the temporal condition. By shifting the ontological hierarchy from temporal to spatial, the axis shifts from horizontal to vertical. Hierarchy assumes a vertical dynamic, where power rests at the top. In Chaos "first of all," there was no hierarchy to account for. "From Chaos came black Night and Erobos. And Night in turn gave birth to Day and Space. Whom she conceived in love to Erobos. And

Earth bore starry Heaven, first, to be an equal to herself . . ."[38] By comparison, the ontology of Zeus's Olympus is mythologically bound up with such verticality; the order he imposes is from above, and depends on such a spatial reckoning of order to perpetuate.

Modernity has enabled and elaborated upon the conditions in which the spatial expansion of an increasingly singular and comprehensive ontological hierarchy occurs. Whereas territorial acquisition was historically a martial pursuit, market economics have changed the nature of imperial expansion. This in not to say market economics have replaced martial territorial expansion, but rather they work together within the ontological framework to expand the spaces in which they assume hierarchical authority. Within the framework of modernity, that hierarchical power structure is reified as a comprehensive system. As Christopher Pendergrast suggested, within the social, economic, and political construct of modernity, the categories of the world are thus governed by the laws of capitalism, and in the process, "the term thus becomes code for closing down alternatives to capitalism."[39] Pendergrast was elaborating on Frederic Jameson's critique of modernity in *A Singular Modernity: Essay on the Ontology of the Present*, in which he observed the comprehensive nature of the system: "The point is that the holders of the opposite position have nowhere to go terminologically."[40] The process of commodification appropriates alternatives and affixes market values onto them, from which profit can be extracted. This interrupts historical progress toward final causes, the subject of a philosophical mode inquiry called teleology.

Teleology is the inherent belief that the course of natural and human events occur within the context of final causes. Plato and Aristotle both explored the idea that history is headed toward a particular conclusion. To Plato, the driving force of teleology was divine; to Aristotle, it was nature itself. It is teleology that provides meaning to history, but what we believe to be at the end of the teleological trajectory defines our understanding of the events that led up to it. Optimism regarding human nature results in certain sympathy for Kronos and his loss in the Titanic War. His captivity in Tartarus is teleology interrupted; the destination of historical directionality is unreachable so long as the Cyclopes and Hekatonkheires keep him down and out.

In his 1784 treatise *Idea for a Universal History from a Cosmopolitan Point of View*, the German philosopher Immanuel Kant argued for a teleology based on reason. In the context of the European Enlightenment, an intellectual movement which sought to discern universal order as something purposeful, meaningful, and rational, Kant wrote that history discerns human actions as a "steady and progressive though slow evolution," and of freedom as an "original endowment."[41] To make this point, Kant articulated nine theses, which build toward his vision of historicity as a process which eventually leads to a "universal cosmopolitan condition." He suggested that universal laws govern the designs and actions of humankind, moving toward an "unknown goal." That there was no "agreed-on plan" with regard to the course of history, humankind was reduced to "folly, childish vanity . . . childish malice and destructiveness." It was up to the philosopher then, to "try and see if he can discover a natural purpose in this idiotic course of things human."[42]

Kant's first thesis was that "all natural capacities of a creature are destined to evolve completely to their natural end." Without reason, Kant argued, the nature of humankind is aimless. Secondly, Kant suggested that reason, a condition that transcends instinct, develops along generational rather than individual lines. The development of reason beyond "an unreckonable series of generations" must be, in Kant's estimation, "the goal of man's efforts."[43] Thirdly, Kant reasoned a philosophical explanation of progressive modernity, suggesting that "nature has willed that man should, by himself, produce everything that goes beyond the mechanical ordering of his animal existence, and that he should partake of no other happiness or perfection that he himself, independently of instinct, has created by his own reason." The development of a comfortable existence through reason, working "up to happiness," was the purpose of human endeavor. Referring to his theory of generational development, Kant posited that "earlier generations appear to carry through their toilsome labor only for the sake of the latter, to prepare for them a foundation on which the later generations could erect the higher edifice which was nature's goal," suggesting this as evidence that human are "a class of rational beings each of whom dies while the species is immortal."[44]

In his fourth thesis, Kant suggested that "antagonism in society" was "the cause of lawful order among men." The social instability that derives from humankind's associations and isolation from others "brings him to conquer his inclination to laziness and, propelled by vainglory, lust for power, and avarice, to achieve a rank among his fellows whom he cannot tolerate but from whom he cannot withdraw." Despite this antagonism, Kant put forth his fifth thesis that "the greatest problem for the human race, to the solution of which Nature drives man, is the achievement of a universal civic society which administers law among men."[45] In Kant's estimation, freedom is the highest purpose of both nature and reason, wherein individual freedom and the freedom of others must be balanced. This problem, as Kant asserts in his sixth thesis, "is the most difficult and the last to be solved by mankind." Kant argued that the sublimation of the individual to a common humanity was necessary for the manifestation of universal freedom. Kant conceded that such development in human reason "will be very late and after many vain attempts."[46]

Just as the individual will had yet to acknowledge a greater good, so too had the particular interests of states prevented the achievement of commonwealth. In his seventh thesis, Kant argued that it was necessary for states to concede their autonomy toward a common good, going so far as to suggest a "league of nations" for such purposes.[47] Such a league would provide a foundation for the institution of a morality that advanced the cause of universal history. Kant's eighth thesis returned to the idea of history as a progression toward commonwealth and "universal cosmopolitanism," specifically stating that "the history of mankind can be seen, in the large, as the realization of Nature's secret plan to bring forth a perfectly constituted state as the only condition in which the capacities of mankind can be fully developed, and also bring forth that external relation among states which is perfectly adequate to this end." Here is some of Kant's biggest use of language, suggesting that "this great revolution" will take a long time to achieve, but is nonetheless inevitable, as "on the fundamental premise of the systematic structure of the cosmos and from the little that has been observed, we can confidently infer the reality of such a revolution."[48]

Kant argued that humanity could hasten the arrival of this shift, through conscientious awareness of historical eventuality and actions which expedite its processes. Among other things, a commitment to public education rather than war was something Kant suggested could "prepare the way for a distant international government for which there is no precedent in world history.".... This gives hope finally that after many reformative revolutions, a universal cosmopolitan condition, which Nature has as her ultimate purpose, will come into being as the womb wherein all the original capacities of the human race can develop."[49] This was something that nations should invest in and pursue, because, according to Kant, it was possible. In his final thesis of the work, he argued that the idea of universal history serves "as a guiding thread for presenting as a system, at least in broad outlines, what would otherwise be a planless conglomeration of human actions." Just as astronomers worked out the plan of the cosmos, so too should historians determine the course of history. By identifying the "guiding thread" of history, "Nature can fully develop and in which the destiny of the race can be fulfilled here on earth."[50]

Taken together, Kant's nine theses in *Idea for a Universal History from a Cosmopolitan Point of View* suggest that history is purposeful, and it is the obligation of each generation to assume the responsibilities of its ultimate purpose, which is to advance the cause of freedom based on the principles of mutuality and commonwealth to a point where it is fundamentally institutional. Without such purpose, Kant wondered "how our descendents will begin to grasp the burden of history we shall leave to them after a few centuries."[51] Ultimately, Kant prioritized the political structuration of universal history as an institutional pursuit, an idea that Georg Freiderich Hegel expanded on as the eighteenth century became the nineteenth and the conditions of modernity became more pervasive.

In 1807, as the Industrial Revolution was beginning to take root in Europe, the German philosopher Georg Wilhelm Friedrich Hegel published *Phenomenology of Spirit*, in which he argued that the fundamental spirit of humanity, its *Geist*, was at its core moral and ethical, and these manifest in the social phenomena of reason and truth.[52] The course of human history was a track of progress toward the realization of a singular ethical truth which he termed "the absolute idea." In *The Philosophy of Right*, Hegel argued that "history is the embodiment of spirit in the form of events," that the *Geist*, or "absolute idea" can only be realized over the course of time, when the "theoretical idea" becomes the "practical idea."[53] More simply put, over the course of time, humanity can transcend its present condition and achieve the absolute idea and make manifest its moral and ethical spirit in a state of nature.

Hegel's most accessible work on the subject of teleology and final causes is *Lectures on the Philosophy of History*, published posthumously in 1837 as a collection of lectures he gave at the University of Berlin between 1821 and 1831. Here, Hegel clearly articulates the concept of teleology and its implications for the way in which we perceive and interpret history. Hegel begins by suggesting that the aim and destiny of history is the *Geist*, or Spirit. According to Hegel's first English translator, J. Sibree, *Geist* is a term which suggests intelligence and more than spirituality in the religious sense of the term. Nevertheless, in spite of having been used "in a rather unusual sense," Sibree concludes that "no term in our metaphysical vocabulary could have been substituted for the more theological one."[54] The realization of Spirit was to Hegel nothing short of the attainment of rational freedom: the liberation from outward control and emancipation from the inward slavery of the lust and passion that causes greed, violence, and destruction.

The course of world history, according to Hegel, "has constituted a rational necessary course of the World Spirit—that Spirit whose nature is always one and the same, but which unfolds this its one nature in the phenomena of the World's existence. This must, as before stated, present itself as the ultimate *result* of History."[55] If everything happens for a reason, what then is the destiny of reason? What is the ultimate design of the world? Hegel asserted that "the History of the world is none other than the progress of the consciousness of Freedom."[56] This progress, in Hegel's view, was limited by the "passions, private aims and satisfaction of selfish desires" that overwhelm the historical record because they "respect none of the aims which justice and morality would impose on them." Unlike Kant, Hegel problematized such antagonism because it prevented the realization of Spirit. We observe injustice through a "retreat into the selfishness that stands on the quiet shore, and thence enjoys the safety the distant spectacle of 'wrecks confusedly hurled.'" The result is nothing short of "history as the slaughter bench at which the happiness of peoples, the wisdom of states, and the virtue of individuals has been victimized."[57]

In identifying the gap between the idea of history as teleological progress and its own reality, Hegel explained that history is a process. While the Spirit exists as a theoretical goal, the "Idea" as practical application of the Spirit historically falls short of realizing it due to short-sightedness and human fallibility. The history of the world, according to Hegel, began "with a general aim—the realization of the Idea of Spirit . . . the whole process of history is directed to rendering this unconscious impulse a conscious one."[58] This is not a clean and straightforward process. Hegel conceded that "the Idea advances to an infinity antithesis" and to comprehend why practice falls short of theory over the course of human history "is the profound task of metaphysics."[59] To bridge this gap, Hegel put a great deal of faith in the idea of the state and the "Great Men" who ran it. Like Kant, Hegel's solution to rectifying the Spirit with the Idea was a political one. Here, Hegel suggested that the absolute and objective Spirit of freedom could be rectified with the means for realizing it, which has historically been a subjective pursuit.

Hegel also considered the temporal and spatial dimensions of his theory. Natural time, according to Hegel, is cyclical: "The changes that take place in Nature—however manifold they may be—exhibit only a perpetually self-repeating cycle." Only in "the region of Spirit does anything new arise." Here humanity breaks out of cyclical time and follows a linear progressive design, which provides for "a *real* capacity for change, and that for the better,—an impulse of *perfectability*."[60] Hegel's conclusion is that "history in general is therefore the development of Spirit in *Time*, as Nature is the development of the Idea in *Space*."[61] To make this case, Hegel explicitly referred to the myth of the Greek Titanomachy to make his point: "Thus, it was first Chronos—Time—that ruled; the Golden Age, without moral products; and what was produced—the offspring of that Chronos—was devoured by it. It was Jupiter [Zeus] . . . that first put a constraint upon Time, and set a bound to its principle of decadence. He is the Political god, who produced a moral work—the State."[62]

Given that Hegel's objective was to argue that the state was the ideal spatial arrangement for the manifestation of linear progressive temporality to manifest, his conventional reading of the Titanomachy is understandable. "Zeus, therefore, who is represented as having put a limit to the devouring agency of Time, and staid this transiency by having established something inherently and independently durable—Zeus and his race are themselves swallowed up, and that by the very power that produced them,—the principle of thought, perception, reasoning, insight derived from rational grounds, and the requirement of such grounds."[63] Hegel was arguing that the state, in its mythological form under Zeus and its technical form under the rule of law, would check the very power that produced it, allowing for a space in which history, as the progress of Spirit, could manifest. The Titans represented cyclical natural time to Hegel, "the dominion of abstract Time, which devours its children," whereas Zeus and the Olympians "embody a spiritual import, and are themselves Spirit."[64]

Where Hegel argued the philosophical construct of human nature and human condition as it manifested in political terms, Karl Marx and Friedrich Engels sought an economic means to achieve teleological ends. Writing a generation after Hegel, their conclusion was that the moral and ethical spirit was inherently socialist. Marx and Engels argued that history was the story of immoral and unethical behavior, exploitation and violence, greed and corruption, and at its root was the economic system of capitalism which promoted competition over cooperation, appropriation over liberation, and individualism over community. Its origins were biblical, as Marx laid out in Chapter Twenty-Six of *Capital*, "The Secret of Primitive Accumulation":

> This primitive accumulation plays in Political Economy about the same part as original sin in theology. Adam bit the apple, and thereupon sin fell on the human race. Its origin is supposed to be explained when it is told as an anecdote of the past. In times long gone by there were two sorts of people; one, the diligent, intelligent, and, above all, frugal elite; the other, lazy rascals, spending their substance, and more, in riotous living. The legend of theological original sin tells us certainly how man came to be condemned to eat his bread in the sweat of his brow; but the history of economic original sin reveals to us that there are people to whom this is by no means essential. Never mind! Thus it came to pass that the former sort accumulated wealth, and the latter sort had at last nothing to sell except their own skins. And from this original sin dates the poverty of the great majority that, despite all its labour, has up to now nothing to sell but itself, and the wealth of the few that increases constantly although they have long ceased to work. Such insipid childishness is every day preached to us in the defence of property.[65]

To undo this systematic oppression of the many by the few, Marx and Engels advocated revolution in their 1848 publication "The Communist Manifesto." After identifying that the means of oppression of the poor, disenfranchised multitude (Proletarians) by the wealthy and privileged few (Bourgeoisie) was the economic system of capitalism, Marx and Engels set forth a series of demands:

1. Abolition of property in land and application of all rents of land to public purposes.
2. A heavy progressive or graduated income tax.
3. Abolition of all rights of inheritance.
4. Confiscation of the property of all emigrants and rebels.
5. Centralisation of credit in the hands of the state, by means of a national bank with State capital and an exclusive monopoly.
6. Centralisation of the means of communication and transport in the hands of the State.
7. Extension of factories and instruments of production owned by the State; the bringing into cultivation of waste-lands, and the improvement of the soil generally in accordance with a common plan.
8. Equal liability of all to work. Establishment of industrial armies, especially for agriculture.
9. Combination of agriculture with manufacturing industries; gradual abolition of all the distinction between town and country by a more equable distribution of the populace over the country.

10. Free education for all children in public schools. Abolition of children's factory labour in its present form. Combination of education with industrial production, &c, &c.

Their belief was that these conditions would bring about the end of class distinctions, and by proxy the dynamics of power and oppression which characterized much of their view of human history.[66] Hegel's *Geist* could only be realized through the destruction of the oppressive systems which kept it a mere theoretical form rather than a practical idea. In a mythological view, this would account for a return to Hesiod's "Golden Age"; in the teleological view, Marx and Engels suggested a temporally progressive series of events that would lead to the liberation of the Proletariat. In either direction, the result is a belief that empowering the disenfranchised, rather than systematically subjugating them to a system which perpetuates itself on inequality, can ultimately bring about social justice. Ultimately, it is an optimistic view of the human condition: equity can be achieved, exploitation and oppression can be abolished, and a moral, ethical and just society can exist.

In Marxist teleology, tribulation comes from below. Marx and Engels famously exhorted: "The Communists disdain to conceal their views and aims. They openly decree that their ends can be attained only by the forcible overthrow of all existing social conditions. Let the ruling classes tremble at a Communistic revolution." Similarly, liberation also comes through struggle from below. The result of the class struggle Marx and Engels argued was inevitable is that "the proletarians have nothing to lose but their chains. They have the world to win."[67] In Marxist timespace, time is the optimistic part of the equation. Marxist theory is imbued with the teleology of liberation. Humanity is heading toward its eventual liberation from the oppressive forces of capitalist markets. Marx does suggest that space is necessary for this evolution to play out. Where that space existed in fifteenth-century Europe, according to Marx, was the commons, where peasants and wage laborers "enjoyed the usufruct of common land, which gave pasture to their cattle, furnished them with timber, fire-wood, turf &c."[68] That space was appropriated by capital interests through the process of enclosures, which Marx argued resulted in "the usurpation of the common lands," and with that the Hesiodic form of temporal degeneracy: "the English working class was precipitated without any transition from its golden into its iron age."[69]

HAUNTOLOGY AND ESCHATOLOGY

Jacques Derrida made a theoretical argument for the phenomenon of "hauntology" in his 1993 book *Specters of Marx*. With regard to the stripping of intrinsic value by commodified use-value, he wrote that: "To haunt does not mean to be present, and it is necessary to introduce haunting into the very construction of a concept. Of every concept, beginning with the concepts of being and time. That is what we would be calling here a hauntology," and concluded that "the commodity thus haunts the thing, its spectre is at work in use-value." The effect of his argument is that modern present is haunted by its own past.[70] The idea of haunting is provocative, and allowed Derrida to consider the haunting of modernity. The material beneficiaries of the modern condition are haunted by those who suffer its inherent inequity. We are bothered to know that the materialism of modernity comes at a human and environmental cost, that the exploitation of the producers by the consumers is ultimately what makes modernity so comfortable for those who inhabit its more exclusive circumstances. Ultimately, we are reminded of the inauguration of Pompeius's theatre in 55 BC, when he had twenty elephants slaughtered to instruct his fellow citizens what it took for Rome to be master of the world. Those in attendance were appalled by the slaughter, cursing Pompeius for exposing them to such brutality. Pompey was simply acquainting them with the means of production, for which "the whole assembly rose up in tears, and showered curses on Pompeius."[71] Modern society is haunted by specters which come to symbolize the acts that produce them. It is a phenomenon that Colin Dickey accounts for in his essay on the aftermath of the housing foreclosure crisis in *The Paris Review*:

> We live among the undead. The things that used to have meaning and purpose—not just houses but banks and governments—have been emptied of what they once meant, and yet they remain, haunting us. We are . . . haunted by forces larger than ourselves, imprisoned by the folly of the rich, who have unleashed some unspeakable dread from which we cannot escape.[72]

The consumer's distance from the means of production and its effects, one of the principle features of modernity, keeps consumers ignorant of the conditions in which their material world is made. The reality of the situation seems almost unfathomable. As Dickey observes, "it seemed impossible that anyone had lived in these places." It is a paradox that Thomas Gray accounted for when he wrote the lines "where ignorance is bliss, 'tis folly to be wise," in England in 1742, the same year the first cotton spinning mill opened in Birmingham, England.

Just as the past haunts the present in hauntological form, so too does that future in eschatological form. It is the stuff of apocalyptic visions, which have proliferated the cultural imagination since the world wars of the twentieth century gave humanity a view of a hellscape that could not be unseen. The fantasies of destruction, playing out in nuclear, environmental, epidemiological, and religious contexts, tell us as much if not more about the time in which they are conceived than they do any imminent or distant future. Modern culture's preoccupation with terminality suggests that it does not believe in its own sustainability. Preparing for the end times has become a daydream for some, a way of life for others, and a business opportunity for others still. Fear, the essence of hauntology, drives the discourse in a temporal direction that gains its own momentum. Jean Beaudrillard's observation in *Simulacra and Simulation* that "the map precedes the territory" evokes the concept of self-fulfilling prophecy.[73] By designing a world in theory, it becomes the world in practice. By imagining an apocalyptic future, we construct the circumstances that lead us to it.

Marx suggested that the specter haunting Europe was "the specter of communism," but his exposition on the history of class struggle in the opening of "The Communist Manifesto" turns the metaphor around to expose the real ghoul: class hierarchy. Put more simply, if Marxist teleology promises deliverance from the evils of capitalism, where, then, did humanity go wrong? The collective mistakes of the past haunt us in the present by reminding us of the possibilities of an alternative reality which only exists in either marginal or theoretical form. Simultaneously, the future haunts the present in the idea of apocalyptic degeneracy. A preoccupation with "the end of the world," an ambiguous phrase usually meaning the breakdown of the modern order, is evident by the pervasiveness of television programs, movies, and books that capitalizes on our collective fascination with eschatology. There is a ubiquitous sense in popular culture that humanity is grinding the gears of the planet, and that an eventual breakdown in the institutions which simultaneously enable and subjugate us is inevitable. Some view this as terrifying, others view it as liberating. The anarcho-primitivism of John Zerzan suggests that "to the question of technology must be added that of civilization itself. Ever-growing documentation of human prehistory as a very long period of largely non-alienated human life stands in stark contrast to the increasingly stark failures of untenable modernity."[74]

Inside of this rhetoric, a fundamental struggle between time and space occurs. The philosophy of Hegel and the history of Marx drive a narrative in which time produces the outcome of human equality and social justice. This outcome is at the expense of the ruling class hierarchy, which has used the discourse of religion and the apparatus of modern circumstance to placate its subjects and enforce its own intrinsic inequality. To prevent the teleology of liberation from occurring *over time*, the nascent and adaptive ruling class hierarchy of the modern world developed and implemented countermeasures that, like Zeus on high Olympus, depend on command and control *over space*. Triangulating, mapping, policing, surveilling, incarcerating, all for the purpose of maintaining an inherently unjust social, political, and economic system that benefits a few at the expense of the many is a state of affairs which requires both historical and geographical inquiry.

Where the teleology of Hegel and Marx is generative and optimistic, that of institutional Christian doctrine and dogma is inherently degenerative and pessimistic. Two primary elements of the Christian religion that have come to underpin much of the history of western civilization, and by proxy, the cultural, political, and economic matrix of the modern world, are the beliefs of original sin and eschatological inevitability. Here, Zeus's reign to control the inherent flaws of the human condition is preferable: control the activities of the monstrous multitude by imposing order from on high, and insist that liberation comes not in this world but the next, all the while amassing fortunes for the few privileged on Olympus at the expense of those "put down" in Tartarus. This is the history of imperialism, racism, militarism, and capitalism.

In the intellectual sense, this eschatological worldview manifests in "declinist theory," which offers the argument that the United States is losing its power and prestige on the world stage, and that this loss is something to be lamented and mourned, while the impending change is something to fear and worry about. Largely a product of post-Cold War foreign policy analysis, scholars like Paul Kennedy, Samuel Huntington, and Robert Gordon articulated a vision of degeneration that referenced political and economic circumstances. Paul Kennedy's 1987 book, *The Rise and Fall of Great Powers: Economic Change and Military Conflict from 1500 to 2000*, argues that the decline of American economic and political might (and by proxy, military power) is an inevitability that must be managed by leaders to ensure "that the relative erosion of the United States' position takes place slowly and smoothly, and is not accelerated by policies which bring merely short-term advantage but longer-term disadvantage."[75] Huntington's 1996 book, *The Clash of Civilization and the Remaking of World Order*, argues that "the west" has been in decline since the turn of the twentieth century, and the rise of "the east" will hasten this demise. Robert Gordon considers the domestic implications of this theory, suggesting that the standard of living for 99% of Americans will remain stagnant or incrementally decline over the course of the next few generations; any improvements will be nominal.[76] Another article which Gordon published in the *Wall Street Journal*, on what was supposed to be the date of the much heralded Mayan apocalypse, applied this idea of decline in economic growth to American innovative capacity.[77] More recently, cultural conditions have factored into the declinist vision, most notably in the work of George Packer's widely received 2013 book, *The Unwinding: An Inner History of New America*, where an absence of morality and "common sense" drives the narrative that America has lost its way and is worse off than generations before.

In an interesting way, declinist theory weaves its way into ideas of American exceptionalism and vice versa. The idea that America is unique in its characteristics and thereby exempt from or immune to history's arcs and patterns in which great powers rise and fall is as old as the idea of America itself. John Winthrop's 1630 sermon arguing that the Massachusetts Bay Colony would be a "city upon a hill" has been quoted by Cold War televangelists, cultural nostalgists, and professional politicians from John F. Kennedy to Ronald Reagan. Reagan added an adjective to make it a "shining city" over the course of his political career, symbolically linking the idea of gleaming material wealth to American exceptionalism at the expense of Winthrop's idea that the example his colony should set was one of Christian charity. Whether they are invoking the idea to stir a crowd into aspirational thinking or lament the loss of the way things used to be, proponents of American exceptionalism often call upon a past that never was, but always will have been. By idealizing the past in terms of "the good old days" or similar nostalgic tropes, a very narrow and exclusive vision of the past emerges which can then be used to frame the present as something deteriorated and ultimately inferior. The idealization of the 1950s as an age of tranquil prosperity, poodle skirts, and jukeboxes is inherently a white male middle- and upper-class fantasy. Those were not "happy days" for anyone else but that privileged few who benefitted from the political, economic, and cultural circumstances which promoted and compounded their interests while internal and external pressures contributed to the "decline" they simultaneously lament.

These two conditions, the original fall of man and the inevitable destruction of humankind, lay at the very foundation the Christian worldview, which has played a large part in constructing the theory and practice of modern capitalism since its formation in the sixteenth century and through the nineteenth century. The pessimism of Christian enterprise, that humankind is corrupt and immoral by its very nature, and that earthly, material progress is neither possible nor practical given the impending end of the world stands in sharp contrast to the ideas of Hegel, Marx, and Engels. It is with this pessimism that the capitalist superstructure builds its prisons, defunds its schools, pursues imperialist wars and extracts resources from the earth for the benefit of a few at the expense of the many. *Fiat voluntas Dei*; God's will be done. As Henry Grattan Guinness wrote in his 1888 work, *The Divine Programme of the World's History*: "History must take one certainly defined course, or else they would be palpably falsified . . . the Bible offers as a pledge of its divine inspiration *a complete programme of future history*."[78]

Ultimately, both Christian and Marxist teleology both present a cycle of liberation through tribulation, although the means are very different. In Christian teleology, the book of Revelation warns of destruction by the hand of God. After the tribulation which lasts thousands of years, God makes "all things new" and ends the suffering on earth. Both tribulation and liberation are granted by authority from above. Ideas about timespace suggest a certain amount of optimism and pessimism with regard to the human condition. In Judeo-Christian timespace, there is an inherent pessimism concerning the concept of time. The Bible suggests an idealized past in the Garden of Eden, and after the fall of man, humanity can only wait until its inevitable, earthly destruction at the hands of God in the book of Revelation. At the same time, the Judeo-Christian worldview is optimistic about the spatial component. God watches over us, sees everything, and angels protect us from peripheral evil. Spatial surveillance by an absolute authority is the primary function of this system.

Historians and metatheorists alike have tried to come to terms with the dual forces of teleology and eschatology in other interesting ways. The twentieth-century historian Arnold Toynbee believed that the teleological bent of history involved the extraterrestrial reconstitution of all living things at some future time:

> Someone who accepts—as I myself do, taking it on trust—the present-day scientific account of the Universe may find it impossible to believe that a living creature, once dead, can come to life again; but, if he did entertain this belief, he would be thinking more 'scientifically' if he thought in the Christian terms of a psychosomatic resurrection than if he thought in the shamanistic terms of a disembodied spirit.[79]

Along those lines, Terence McKenna believed that chaos is merely the inability of the human mind to perceive larger forces at work in the context of universal unity, and that over time, this order manifests fully, creating a new higher order in the process.[80] In *The Invisible Landscape*, McKenna argued for the eventual manifestation of the Eschaton, facilitated by a mathematical calculation based on the permutations of the *I Ching* which, when compared to a graph charting the course of world history, predicted the occurrence of "Time Wave Zero," where he predicted an end to the novelty of history transcendence into a higher plane of existence.

In contrast, Anthony Giddens altogether rejects the idea that history is evolutionary or that change is an elementary factor in measuring time. In *A Contemporary Critique of Historical Materialism*, he writes that associating time with social change is "an elementary, though very consequential, error." One aspect of Giddens's reasoning for this assertion is that time is "as necessary a component of social stability as it is for change."[81] Two of the aspects of stability Giddens explores are tradition and routinization; tradition being the manifestation of ontological security and routinization being the economic commodification of tradition by which it becomes stripped of moral meaning.[82] Both of these concepts fail to acknowledge the evolution of tradition and/or routine, therefore negating his argument that change is a primary element in the measurement of time.

The geographer Yi-Fu Tuan suggests that the connection between time and space is experience. According to Tuan, experience is the way by which we organize time and construct our understandings of space and place. Giddens articulates three intersecting planes of temporality: the *durée*, which is the temporality of immediate experience; the *dasein*, which is the life cycle of an organism; and the *longue durée*, which is long-term institutional or historical time.[83] Yi-Fu Tuan complicates the concept of institutional or historical time by suggesting three types: Cosmogenic, Human, and Astronomical. The first two are linear and one directional, the last one is cyclical and repetitious.[84] Whether or not time is evolutionary or revolutionary then becomes the question. Where Giddens suggests human history neither evolutionary nor directional, Tuan suggests that linear and one-directional time exists within a circular and cyclical frame. Tuan's articulation of time is evocative of William Butler Yeats's poem "The Winding Stair," in which events recur in human history at higher levels than previously.[85] Regardless of its form, change is a primary factor in temporal conceptualization.

Where Hegelian and Marxist teleology is inherently liberating through the agency of humankind to make this physical, material world a place of social justice and equality, Christian eschatology is convinced of humankind's inherent decrepitude and ultimate destruction; liberation comes only in death. Teleology is liberation from below, eschatology is control from above. The apocalyptic fervor which coincides with the rise in the social, political, and economic systems of modernity from 1500 to 1800 suggests a rhetoric created to explain the harsh realities of the plantation, the factory, the pit, the ship, and the prison. Rather than justify those conditions, the rise of revolutionary ideology during that pivotal time in human history suggests a turn toward teleological hope for something better. Blake's description of Satanic Mills was not a justification for the suffering of the working classes; it was a condemnation of the conditions imposed on them from above, and a challenge to the system that leveled them.

Challenging the foregone conclusion of the end of days and the degenerative temporal cycle that went with it meant challenging the religious worldview that had defined the western worldview until that time. As Arthur Williamson writes, "for most of its history, European culture had visualized time and change as marginal, irrational, and emblematic of man's fallen state. The burden of the western message has been atemporal and indeed anti-temporal." In the meantime, a system of spatially ordered hierarchical command and control had evolved. But it was vulnerable. From beyond and below the peripheral borderlands and the institutions of control and exploitation within the kingly domains, a fundamental challenge to that hierarchy emerged. Williamson continues, "the apocalypse underwrote the Reformation in the sixteenth century, the British Revolution in the seventeenth century, and the American Revolution in the eighteenth century." He continues that "before anything else, the apocalypse and its attendant complex of ideas comprise mechanisms for imagining time."[86] Once doomed by Christian doctrine and the systems that enforced it, new ways of thinking about human nature and the human condition appeared from the margins and arose from the depths to challenge the fundamental order of things, only to be challenged by new and formidable manifestations of Cyclopes and Hekatonkheires. The history of the modern world is every bit the history of the struggle between the teleological liberation and the eschatological destruction of the material world.

Both teleology and eschatology are linear and directional; the former is generative and constructive, the latter is degenerative and destructive. The result of these oppositional forces is an ontological stasis, enforced by hierarchical ordering of time and space, to prevent anything from happening that would push the direction of timespace toward teleological liberation. With the teleological threat to hierarchical order bound in chains behind walls monitored by its monsters, the Olympian order is free to engage its degenerative, eschatological pursuits. This is the condition of modernity; the apparent feedback loop of *plus ça change, plus c'est la même chose,* the superficial remodeling of the material condition without any substantial reconfiguring of the social, political, or economic relations in which they manifest. All the while, those social, political, and economic relations grind toward the deterioration of the human condition and the material world. Timespace therefore becomes the metaphysical battlefield upon which the elemental forces of production and destruction do battle.

This modern manifestation of the Titanomachy plays out over the course of history in the revolutions, the mines, the libraries, and the office buildings, and everywhere in between. Its battleground expands over natural and built landscapes. We are all participants, enforcing Zeus's hierarchical ordering or undermining it, locking and chaining the army of Kronos behind walls or plotting with Ananke to unseat Zeus's order, and establish a new order where commonwealth and social justice takes precedence over the commodified order of private property and the accumulation of material wealth. They are ideals we are innately aware of, but the systemic and encompassing nature of ontological stasis promotes a certain resigned pragmatism which breeds resignation from the battle. It is too hard to resist. It is easier to concede. With each concession, the process of eschatological degeneration continues.

The opinion of whether the world of Kronos or the world of Zeus is preferable is based largely on one's beliefs regarding the fundamental nature of the human condition. Myth, according to the eighteenth-century political philosopher Giambattista Vico, who wrote *The New Science* in 1725, suggested that "the first fables must have contained civil truths, and must therefore have been the histories of the first peoples."[87] Thus, the myth of the Titanomachy illuminates fundamental questions about human nature and how to come to terms with it. Kronos, representing the elemental construct of time, introduces the concept of teleology, represented by his partner Ananke, the goddess of inevitability. Zeus and the Olympians enforce their hierarchical order from above and, as Plato's Eleatic visitor put it: "conduct their defence from some invisible height." Modernity is both the time and place where the most recent battle between these forces manifests. To borrow from the labor activist Florence Reece, who put the question to verse while her home was under siege by Harlan County Sheriffs at the behest of the mining company where her husband was organizing the workers to unionize in 1931: "Which side are you on?"[88]

TAKING TIME AND SPACE

Modernity has a profound effect on our constructions and perceptions of time and space. The British sociologist Anthony Giddens insists that the most defining property of modernity is that we are disembedded from time and space, and that the abstractability of time and space are a fundamentally unique feature of modernity.[89] This phenomenon, which allows social systems to expand their reach and exercise power across increasingly large expanses of space and time, is something Giddens refers to as "time-space distanciation." In premodern societies, time and space could only be experienced from a peripatetic, embodied perspective. In modernity, we can abstract time and space and experience them by proxy. In *The Great Accelerator*, the French cultural theorist Paul Virilio defines modernity as speed, qualifying our preoccupation with it as a "technocult." Speed challenges spatial distance through temporal means. Derrida's "The Postcard" demonstrates how much modernity has to do with distance, both in physical and ideological terms; both of which are ultimately spatial considerations.[90] Time and space are resonant themes in academic considerations of modernity.

The geographer David Harvey argues the result of these developments is a phenomenon he calls "time space compression." In *The Condition of Postmodernity*, Harvey outlines "processes that so revolutionize the objective qualities of space and time that we are forced to alter, sometimes in quite radical ways, how we represent the world to ourselves." He invokes the term compression "because a strong case can be made that the history of capitalism has been characterized by speed-up in the pace of life, while so overcoming spatial barriers that the world sometimes seems to collapse inwards upon us."[91] The forces of modernity make for a certain omnipresent imminence within the framework of the human condition: everything, all the time, everywhere. The ubiquity of the global forces which drive the modern condition is new, and something that humanity has yet to fully come to terms with.

Barney Warf explores the theoretical history of this phenomenon in his 2008 book, *Time-Space Compression: Historical Geographies*. Organized temporally, beginning with the "three broad world-historical epochs of capitalism: The early modern, late modern and postmodern," Warf asserts that "the current epoch is but the latest chapter in a long series through which the world's cultures have given time and space a dramatically new texture." That the present condition has come to pass is not, however, something that Warf accepts as a teleological eventuality. Rather, each epoch "was the contingent, unintentional outcome of its predecessor."[92] Time-space compression was not inevitable by this construction, but had as much to do with the development of capitalism from its mercantile and colonial origins in the early modern epoch toward its industrialized imperialism in the late modern period and ultimately the globalized transnational neoliberal economic order of the much debated characterization of a postmodern present.

The central focus over the course of this book is to understand how the forces of modernity, broadly conceived and with respect to the ideological debate which follows the term, have come to shape our perceptions of the time and space in which it manifests. Time and space have both natural and humanly defined boundaries. An example of how modernity affects our perceptions of time and space is the manner with which, over the course of modern history, socioeconomic constructions of temporality have eclipsed the natural. Today, the humanly defined boundaries of time, the clock and the calendar, have come to obfuscate natural temporal cycles. Day and night, the seasons, the solar and lunar cycles, are all obliterated by the socioeconomic forces which drive the technological development of modernity. This is to say nothing of our connection to the galactic cycle and our place in the universe as a whole. The result is a complete disconnection with the natural cycles of time. Markets demand twenty-four hour access to goods and services and technology enables it. The language of time has become itself economic: we spend, save, and waste time, and increasingly little of it is "free." We obey the alarm clock on a Monday morning despite our innate reasoning to do otherwise. Time becomes a tool of control rather than liberation.

Similarly, modernity has implications for our understanding and conception of space. Technology has triangulated and obliterated natural topographical boundaries; we traverse the planet irrespective of mountain ranges, bodies of water, and changes in landscape. Social boundaries resulting from ground-level human interaction, neighborhoods, communities, and regions, have given way to boundaries imposed by bureaucratic authority from the top down: school and voting districts; commercial, residential, and agricultural zones; counties, states, and nations. Unlike natural boundaries which are often comprised of physical and/or social borderlands, fluid and subject to change dependent on natural and social conditions, modern borderlines are exacting, linear, and precise. They are often overtly political, often at the expense of the economic and cultural conditions which preceded them. The local gives way to the global, the democratic gives way to the authoritarian, and the social concedes to the political and economic.

Just as time-space compression has political, economic, and social implications, so too does resistance to it. From mass protests and boycotts to personal lifestyle choices, reactions to the conditions of time-space compression are myriad. Antiglobalization demonstrations, beginning most prominently in Seattle in 1999, show a broad-based resistance to the ordering of global capitalism that takes into consideration environmental and labor issues along with the rights of citizens to engage society on an equal footing. Encouragement to "buy local," and to "support small business" reflects an ethos that takes into account the quality of life not only of the consumer, but also the producer, and the immediate environment in which they both live. In doing so, it also takes into account the quality of life of the community in which the transaction takes place; more of a commonwealth arrangement than one of imperialist, commercialized extraction. Even exhortations to "take time," as evidenced by the growth of the Slow Food Movement and the abandonment of automobility in exchange for walkable, bikeable communities are evidence that the conditions of time-space compression are not always ideal. In addition to taking time, new forms of resistance focused on "taking space" are extending the challenge further. The Occupy Movement, which began in 2011, resonated for its simplicity and effectiveness. Simply being somewhere for a period of time can be an act of resistance within the framework of modernity.

NOTES

1. Hesiod, *Theogony/Works and Days*, trans. M. L. West (Oxford: Oxford University Press, 1988), 40–42.
2. Joseph Campbell, *The Hero with A Thousand Faces* (New York: New World Library, 2008), 224–227.
3. The Venerable Bede, *The Reckoning of Time*, trans. Fait Wallis (Liverpool: Liverpool University Press, 1999), 158.
4. Ibid. 249.
5. Peter N. Stearns, "A Cease Fire for History." *The History Teacher*, Vol. 30, No. 1 (Nov. 1996): 70–71.
6. Ibid. 67, 81.
7. Ernst Breisach, *Historiography: Ancient, Medieval and Modern, Third Edition* (Chicago: University of Chicago Press, 2007), 1–5.
8. Cassiodorus, *The Letters of Cassiodorus*, trans. Thomas Hodgkin (London: Henrey Frowde, 1886), Book III, Letter 5, 200.
9. Theodore Mommsen, "Petrarch's Conception of 'The Dark Ages.'" *Speculum*, Vol. 17, No. 2 (April 1942): 241–242.
10. Leonardo Bruni, *History of the Florentine Peoples*, trans. James Hankins (Boston: Harvard University Press, 2001), Vol. I, Books I–IV, xvii.
11. Ernst Breisach, *Historiography: Ancient, Medieval and Modern, Third Edition* (Chicago: University of Chicago Press, 1983, 2007), 181.
12. United Nations Population Fund, "Urbanization: A Majority in Cities," Last modified May, 2007. www.unfpa.org/pds/urbanization.htm
13. E. H. Carr, *What Is History?* (New York: Penguin Books, 2008), 23.
14. Yi-Fu Tuan, *Space and Place: An Experiential Perspective* (St. Paul: University of Minnesota Press, 1977), 148.
15. Peter Gay, *The Enlightenment: The Science of Freedom* (New York: Norton, 1977), 2.
16. Michael Lind, "The Age of Turboparalysis," *Salon*, Last modified Dec. 27, 2011, www.salon.com/2011/12/27/the_age_of_turboparalysis/
17. Goody, *The Theft of History* (Cambridge: Cambridge University Press, 2006), 161.
18. Ibid. 129.
19. Peter Van Der Veer, "The Global History of 'Modernity.'" *Journal of Economic and Social History of the Orient*, Vol. 41, No. 3 (1998): 287.
20. C. A. Bayly, *The Birth of the Modern World, 1780–1914* (Malden: Blackwell, 2004), 10–11.
21. William Everdell, *The First Moderns: Profiles in the Origins of Twentieth Century Thought* (Chicago: University of Chicago Press, 1997), 9.
22. Jose C. Moya, "The Massification of International Families in the Nineteenth Century," in *Transregional and Transnational Families in Europe and Beyond: Experiences Since the Middle Ages*, ed. C. Johnson (New York: Berghahn Books, 2011), 32.
23. Gordon E. Moore, "Cramming More Components onto Integrated Circuits." *Electronics*, Vol. 38, No. 8 (April 19, 1965): 4.
24. Ray Kurzweil, "The Law of Accelerating Returns," Last Modified March 7, 2001. www.kurzweilai.net/the-law-of-accelerating-returns

25. Anthony Giddens, *Modernity and Self Identity: Self and Society in the Late Modern Age* (Stanford: Stanford University Press, 1991), 16.
26. Alvin Toffler, *Future Shock* (New York: Random House, 1970), 15.
27. Douglass Rushkoff, *Present Shock: When Everything Happens Now* (New York: Current Press, 2013), 243.
28. Tristan Vey, "Welcome to the Anthropocene, Earth's New Era," *Worldcrunch*, Last modified October 26, 2011, www.worldcrunch.com/welcome-anthropocene-earth-s-new-era/tech-science/welcome-to-anthropocene-earth-s-new-era/c4s3982/
29. Immanuel Kant, "Prolegomena to Any Future Metaphysics," in *Classics of Western Philosophy*, ed. Steven M. Cahn (Indianapolis: Hackett Publishing, 1990), 944.
30. Shailer Mathews, *The Spiritual Interpretation of History* (Cambridge: Harvard University Press, 1920), 31.
31. Ibid, 34–35.
32. Hayden White, *Metahistory: The Historical Imagination of Nineteenth Century Europe* (Baltimore: Johns Hopkins University Press, 1973), 426–428.
33. Martin W. Lewis and Karen E. Wigen, *The Myth of Continents: A Critique of Metageography* (Berkeley, Los Angeles, London: University of California Press, 1997), ix.
34. Julian Cowley, "Free Your Mind Your Ass Will Follow." *The Wire*, No. 295, September 2008, 40.
35. Benjamin Franklin, "The Morals of Chess," in *The Chess Player*, ed. George Walker Teacher (Boston: N. Dearborn, 1841), 8.
36. Plato, *The Sophist and the Statesman*, trans. A. E. Taylor (London: Thomas Nelson and Sons, 1961), 246 A–C, 143.
37. Reinhardt Grossman, *The Existence of the World: An Introduction to Ontology*, (London: Routledge, 1992), 1.
38. Hesiod, "Works and Days," in *Hesiod and Theogonis*, trans. Dorothea Wender (New York: Penguin Books, 1973), 26.
39. Christopher Pendergrast, "Codeword Modernity." *New Left Review*, Vol. 24 (November-December 2003), 24.
40. Frederic Jameson, *A Singular Modernity: Essay on the Ontology of the Present*, (London: Verso, 2002), 10.
41. Immanuel Kant, "Idea for a Universal History from a Cosmopolitan Point of View," in *Immanuel Kant, On History*, trans. Lewis White Beck (New York: Bobbs-Merrill Co., 1963), 11.
42. Ibid. 12.
43. Ibid. 13.
44. Ibid. 14.
45. Ibid. 16.
46. Ibid. 17–18.
47. Ibid. 19.
48. Ibid. 22.
49. Ibid. 23.
50. Ibid. 25.
51. Ibid. 25.
52. Georg.Wilhelm Friedrich Hegel, *The Phenomenology of Spirit*, trans. A. V. Miller (Oxford: Clarendon Press, 1977), 263–266.
53. Georg Wilhelm Friedrich Hegel, *The Philosophy of Right*, trans. S. W. Dyde, (Kitchener: Batoche Books, 2001), 270.
54. Georg Wilhelm Friedrich Hegel, *Lectures on the Philosophy of History*, (London: G. Bell and Sons, 1910), iv.

55. Ibid. 11.
56. Ibid. 19–20.
57. Ibid. 22.
58. Ibid. 26.
59. Ibid. 27.
60. Ibid. 56.
61. Ibid. 75.
62. Ibid. 79.
63. Ibid. 80.
64. Ibid. 254.
65. Karl Marx, *Capital: A Critical Analysis of Capitalist Production* (London: Swan & Sonnenschein & Co., 1902), 736–737.
66. Karl Marx and Freiderich Engels, *The Communist Manifesto* (New York: Bedford/St. Martins, 1999), II: Proletarians and Communists.
67. Ibid. 22.
68. Karl Marx, *Capital: A Critical Analysis of Capitalist Production* (London: Swan & Sonnenschein & Co., 1902), 740.
69. Ibid. 741–742.
70. Jacques Derrida, *Specters of Marx* (New York: Routledge, 1994), 10.
71. Pliny, *The Natural History of Pliny*, trans. John Bostock and H. T. Riley (London: Henry G. Bohn, 1855), Vol. II, Book VIII, 254.
72. Colin Dickey, "Unhousing," *The Paris Review*, Last modified March 19, 2014, www.theparisreview.org/blog/2014/03/19/unhousing/
73. Jean Baudrillard, *Simulacra and Simulation*, trans. Sheila Faria Glaser (Ann Arbor: University of Michigan Press, 1994), 1.
74. John Zerzan, "Why Primitivism?" Last modified, 2002, www.johnzerzan.net/articles/why-primitivism.html
75. Paul Kennedy, *The Rise and Fall of Great Powers* (New York: Vintage Books, 1987), 534.
76. Robert J. Gordon, "Is U.S. Economic Growth Over? Faltering Innovation Confronts The Six Headwinds, NBER Working Paper No. 18315, Issued in August 2012. www.nber.org/papers/w18315
77. Robert J. Gordon, "Why Innovation Won't Save Us," *Wall Street Journal*, Dec. 21, 2012.
78. Henry Grattan Guinness, *The Divine Programme of the World's History* (New York: Hodder and Stoughton, 1888), 445–446.
79. Arnold Toynbee, *Experiences* (Oxford: Oxford University Press, 1969), 141–142.
80. Terence McKenna, *The Invisible Landscape: Mind, Hallucinogens and the I-Ching* (New York: Seabury, 1975), 193–201.
81. Anthony Giddens, *A Contemporary Critique of Historical Materialism* (Berkeley and Los Angeles: University of California Press, 1983), 17.
82. Ibid. 152–154.
83. Ibid. 19–20.
84. Yi-Fu Tuan, *Space and Place: An Experiential Perspective* (Minneapolis: Minnesota University Press, 1977), 131.
85. David Punter, *Modernity* (New York: Palgrave Macmillan, 2007), 2.
86. Arthur Williamson, *Apocalypse Then: Prophecy and the Making of the Modern World* (Westport: Praeger Publishers, 2008), 2–3.
87. Giambattista Vico, *The New Science of Giambattista Vico*, trans. Thomas Goddard Bergin and Max Harold Fisch (Ithaca: Cornell University Press, 1948), 65.
88. Florence Reece, "Which Side Are You on?" in *Here Comes A Wind: Labor On The Move* (Chapel Hill: Institute for Southern Studies, 1976), Vol. 4, 1–2, 90.

89. Anthony Giddens, *Modernity and Self Identity: Self and Society in the Late Modern Age* (Stanford: Stanford University Press, 1991), 2–18.
90. David Punter, *Modernity* (New York: Palgrave Macmillan, 2007), 74–78.
91. David Harvey, *The Condition of Postmodernity: An Enquiry into the Origins of Cultural Change* (Cambridge, Blackwell: 1989), 240.
92. Barney Warf, *Time-Space Compression: Historical Geographies* (New York: Routledge, 2008), 10–12.

2 Modern Timespace

INTRODUCTION

In the late fourth century AD, Saint Augustine struggled to articulate a basic definition of time: "What then is time? If no one asks me, I know; if I wish to explain it to him who asks, I know not."[1] Writing in his *Confessions*, Augustine attempted to express the origins of timespace in theological terms. He concluded that time is something distinct and separate from space, yet we depend on the movement of bodies in space to measure time. Augustine problematized this accordingly: "let no man tell me that the motions of the heavenly bodies are times."[2] Ultimately, Augustine concluded that time was an invention of the human mind; that remembering evoked the past, expectation proposed a future and consideration informed the present. In this construction, past and future can only be thought of as present in the forms they manifest: memories and foretelling.[3]

Augustine's reference to planetary motion begins to expose the margins of temporal conditions. The natural boundaries of time are somewhat cyclical: day and night, the seasons, the solar and lunar cycles, and even galactic cycles are factors in long-form temporal expression. These forms of time are largely variable. Like Augustine conceded, the conceptualization of time necessitates spatial estimation. The variability of these conditions exposes the relationship between time as a philosophical concept and the physical conditions of the natural world. Days are longer in the summer and shorter in the winter. Seasons are theoretically marked by solstices and equinoxes, but their weather patterns can be somewhat erratic. Years are marked by the earth's elliptical circumnavigation of the sun, which overlaps with moon phases and planetary cycles. Understanding natural time means understanding our connection to the universe; something that the conditions of modernity, and the constructions of time in which they manifest, most notably the calendar and the clock, often tend to obfuscate.

Like time, space also has natural and humanly defined boundaries. The natural boundaries of geographical space are environmental, and include mountains, valleys, bodies of water, and tree lines, along with transitional spaces between ecosystems like the North African Sahel and the Canadian

Shield. These boundaries are typically variable and meandering. The geographical relation of humanly defined space, since Hesiod recounted Zeus's Cyclopes and Hekatonkheires to defend the Olympian ordering from Chaos, is predicated on maintaining power through enforcing ontological stasis, and preventing the teleological progression of history from occurring. Over the course of history, the triangulation of geographical space has often served to control, challenge, or altogether prevent the dynamics of historical change. As the triangulation of space has become more exacting with the onset of modernity and its exponentially rapid shift from *mythos* to *techne*, the dynamic of teleological change has also become increasingly difficult to assert and to realize in surveilled space. When properly monitored and enforced from above, space is static; it prevents unregulated change. The Cyclopes and Hekatonkheires protect high Olympus from Typhon's monsters who would unseat Zeus and return the world to Kronos and "The Golden Age."

With Zeus's victory over the Titans, Kronos and his cohorts were cast *away*. This is to say that they were spatially removed from power, from the Olympian center to the Chaotic periphery. With this event began the timescape of Greek mythology from which classical history emerges. From the physical ordering of timespace within the foundational creation stories of the cultural construction, which its proponents call "Western Civilization," comes modern social, political, and economic order. Order gives rise to a hierarchy with the responsibility of maintaining it, lest the metaphoric, built structure collapse, leaving nothing but primeval and elemental Chaos. Within these structures, the vilification of revolutionary ideology as a manifestation of Chaos (which inherently threatens order) is a matter of course. The preservation of order becomes the highest function of its attendant institutions.

The origins of the term "timespace" are in both physics and philosophy. Physicists have been using the term "time-space" and "space-time" somewhat interchangeably since the turn of the twentieth century to consider matters pertaining to the theoretical nature of their discipline. The term was first explicitly proposed by Hermann Minkowski in his 1908 essay "Raum und Zeit," though its theoretical origins lie as far back as Jean D'Alembert's article "Dimension" in Volume 4 of *Encyclopdie* published in 1751.[4] The use of the term in physics suggests the combination of space and time into one mathematical continuum. The use of the term "timespace" in the philosophical sense is a product of postmodernist studies dating back to the 1980s, when social scientists began using the term to consider the works of Marx and Engels, Kant, and Hegel in revisionist terms. Anthony Giddens qualifies his exposition of the term as a "neologism" in his 1983 work, *A Contemporary Critique of Historical Materialism*.[5] Philosophers and geographers quickly adopted the term, and began applying it to postmodern theory not long after.

The formalized philosophical ideas behind the term "timespace" were forming around the same time that physicists came to explore the concept at the turn of the twentieth century. The idea that time as a measure of change manifests in spatial terms was first articulated by Martin Heidigger in his 1927 work, *Being and Time*, in which he suggested that time and space are a singular phenomenon in contrast to the classical, Euclidean view that they are only manifest in things and events. With regard to the concept of the ontological nature of existence, Heidegger wrote that "this being that we call Da-sein must be addressed co-ordinately 'as temporal' 'and also' as spatial."[6] Time and space are intrinsically connected in physics and in philosophy. Time needs space to manifest in, and spatial conditions develop over time.

If people have a universal interest in autonomy and independence from systems that impinge on their fundamental rights as humans being, something Hegel called "the phenomenology of spirit," what then has prevented them from asserting themselves to those ends over the course of historical events?[7] In the Titanomachy, Zeus's autocratic and hierarchical command of Olympian space was imposed to preserve order. With the help of the Cyclopes, who surveyed the space and imposed masonic order upon it by building walls which increasingly marginalized and excluded the Titanic resistance, the Olympian hierarchy rendered time static and interrupted the teleology of emancipation. This ancient pattern continues in the context of modernity through the imposition of systems which order and designate space to impose halting stasis, rendering it rational and abstract. Three of these systems, technological development, the dual forces of urbanization and nationalism, and capitalist market economics, undermine a fourth persistent theme of the human condition within the context of modernity: the presence of a democratic ideology. This latter theme is a manifestation of emancipatory phenomenology, and therefore remains peripheral and insurgent: Kronos against Olympus.

The static geographical model differs from the dynamic temporal model in that the former rests its authority on the pretense of order and constancy, whereas the latter challenges that order with a chaotic variability hedging toward a teleological end. Henri Lefebvre observed that space is produced to reinforce political, economic, and cultural hegemony.[8] While geography certainly accounts for change within its own value system of proprietary authority (borders, infrastructure, natural resources, populations, etc.), throughout modernity it has been used as a tool of control to manage, and at times even to prevent, the possibility of historical change that would undermine centralized control of those systems.

The origins of the struggle between centralized, hierarchical spatial authority and the temporal challenge to it from the periphery were set eons ago in the mists of mythological time. Their application in social, political, and economic affairs has been the course of recorded human history. While a comprehensive account of every instance on the historical record which substantiates this formula would be an impossibly exhaustive exercise, a thematic consideration that takes each of the components of modernity—technology, the dual forces of nationalism and urbanization, market economies, and the presence of democratic ideologies—and considers the effect of each on the human constructions of time and space, begins to offer a sense of the comprehensive nature of modern timespace.

TECHNOLOGY AND TIME

Modern forms of technology have obliterated humanity's intuitive perception of time. The effect is comprehensive. The mechanization of time in the early modern era resulted in a preoccupation with precision and speed that obfuscates natural rhythms, cycles, and patterns. Moments, whiles, spells, periods, and stretches of time become both impractical and implausible. Mechanical constancy replaces natural variation. The result is an increase in the speed of modern life, without any sense of direction.

Prior to the widespread adoption of mechanical timekeeping for industrial organization in the nineteenth century, concepts of time were largely linked to natural phenomena. Clepsydrae, or water clocks, were commonly used to mark units of time throughout the ancient world in Babylon, Egypt, India, China, and Greece. These short-term temporal indicators were variable, and for the most part, the day was the predominant measure of time. Within the constant twenty-four hour cycle though, the variable of sunlight depending on the season dictated the patterns and rhythms of daily life. Sundials told time, measuring the solar passage from rise to set. These patterns were largely set by agricultural work rhythms, which depended entirely on the diurnal pattern of the sun and the annual pattern of the seasons.

During the medieval period, the variability of temporal experience began to become regulated and systematized. The medieval *horae* took natural time into account. The most basic unit of time was still the day, not the hour, and the canonical hours observed by religious orders were variable. Gerhard Dorn Van-Rossum observes in *The History of the Hour: Clocks and Modern Temporal Orders* that "in medieval understanding, too, the division of time, and especially of the day, was not simply a given fact, beyond doubt and unchangeable. Instead, it was seen as determined in part by natural rhythms, in part by social convention or 'political decisions,' and as subject to historical change."[9] Those natural rhythms connected the physical world to the spiritual world in a cohesive and comprehensive manner.

With the spread of Renaissance Humanism as the medieval transitioned to the modern, the French Calvinist scholar Joseph Justus Scaliger coined the term *chronologia* in the sixteenth century to describe the new science of calendrical calculation, and used the term *computus* to describe the process of mathematical reasoning behind his idea. Scaliger conceived time as the range of "celestial motion," and argued that modern chronologia, or timekeeping, must be based on astronomical observation.[10] Temporal reckoning was originally intended to enhance humanity's understanding of cosmological cycles. Since Scaliger's initial exposition, the advance of temporal reckoning as a technological exercise has gone far beyond the human perception of what time actually is. Today the Ytterbium atomic clock at the National Institute of Standards and Technology in Boulder, Colorado measures time with an accuracy of two parts in one quintillion.[11] It is a clock designed to aid in the development of machines, not people.

At the same time, this technological pursuit of temporal accuracy has isolated humanity from those natural cycles Scaliger sought to unite it within the sixteenth century. Most people are unaware what phase of the moon it is at any given time. There is no practical need for such knowledge, as technology insulates us from the natural world. We become detached and unfamiliar with its patterns and cycles, in a word: desensitized. Richard Brautigan reflected on the unrealized liberating potential of technology in his 1967 poem "All Watched Over By Machines of Loving Grace," imagining a time when the capacity of machinery would uplift humanity to a condition "where we are free of our labors and joined back to nature."[12] In doing so, Brautigan echoed the sentiment of Okura Nagatsune, who wrote the *Nôgu-benri-ron* (*On the Efficiency of Farm Tools*) toward the end of the Tokugawa period in early modern Japan, with the primary aim of "reducing the peoples' labor."[13] Technology, in both Brautigan's and Nagutsune's views, could have helped connect rather than isolate humanity from nature by freeing rather than preoccupying our time.

Technological development as it relates to the physical phenomenon of speed has also had a profound effect on modern temporal reckoning. For the course of human history, the speed of communication was limited to the physical speed of transportation systems: roads and horsepower on land, and the wind and waves at sea. With the development of electrical transmission systems, the speed of communication outpaced the speed of human mobility. Ideas move faster than people do. The shift from the transportation model of communication to the transmission model has precursors in the ancient and medieval worlds, with the existence of talking drums, smoke signals, heliography, and torch signals, but the introduction of electricity allowed humans to communicate at a distance beyond visual and auditory perception.

The introduction of the telegraph in the 1840s in the United States and Europe allowed for previously unprecedented time-space distancing in communications. As Brooke Hindle wrote in *Emulation and Invention*: "The electromagnetic telegraph, like the steamboat, was something new under the sun. It was not an improvement within an existing technology; and it was not a combination of existing capabilities put together to answer a clear social need."[14] The first transatlantic cable, laid between Ireland and Newfoundland in 1858, was a first step toward globalizing electronic communication patterns which maps of internet traffic today show is still predominant. The telephone and internet communications are essentially variations on the technological model of the telegraph: they are more refined and developed versions of communication through electrical transmission. Their ubiquity is evidence of the proliferation of technology in the modern world.

As with communications, technology has had a profound influence on time as it relates to transportation. Here again, speed is the primary objective and function of technological development as it relates to transportation in the modern era. Prior to the advent of the steam powered engine in 1781, horsepower was a literal term. People and things moved at speeds dictated by biology and geography. The application of the coal powered steam engine, and later the oil and gas powered internal combustion engine to transportation had a profound effect on the modern world as it developed in the late nineteenth and early twentieth centuries. Railroads and ships made transportation faster, and the increased efficiency was a primary interest of the commercial markets the machines served. In 1903, the Wright Brothers flew a fixed wing aircraft powered by a gasoline engine for a distance of 120 feet at a speed of 6.8 miles per hour, ten feet off the ground. Sixty-six years later, NASA astronauts flew the Apollo spacecraft a distance of just under one million miles at a peak speed of 24,000 miles per hour, over the course of four days, to land human beings on the moon. Automobility is marketed as the democratization of speed for those who can afford it. Milan Kundera wrote in his 1997 novel, *Slowness*, that "speed is the form of ecstasy the technical revolution has bestowed on man. As opposed to a motorcyclist, the runner is always present in his body, forever required to think about his blisters, his exhaustion; when he runs he feels his weight, his age, more conscious than ever of himself and of his time of life. This all changes when man delegates the faculty of speed to a machine: from then on, his own body is outside the process, and he gives over to a speed that is noncorporeal, nonmaterial, pure speed, speed itself, ecstasy speed."[15]

Speed has also been a defining factor in the technological development of modern warfare. The introduction of mechanized vehicles to combat in World War I, most notably the airplane and the tank, sped the pace of battle as never before. Speed has always been a primary factor in the history of warfare, but prior to industrial mechanization it has been limited, like civil transportation, by biology and geography. General Schleiffen's two-front plan for Germany in World War I was predicated on speed and industrial might. Speed was also the linchpin of the German blitzkrieg strategy of World War II, which was predicated on densely concentrated motorized armor and air power to break through and encircle an enemy. General Hans von Seeckt's idea of using the overwhelming speed and power of modern weaponry to stun and overtake an enemy was applied to post-Cold War military strategy in 1997 by Harlan K. Ullman and James P. Wade Jr. in a paper entitled "Rapid Dominance—A Force For All Seasons. Technologies and Systems for Achieving Shock and Awe: A Real Revolution in Military Affairs." The authors noted that "rapid Dominance is defined by four chief characteristics: total knowledge of self, environment and the adversary; rapidity; brilliance in execution; and control of the environment."[16]

Ultimately, the effect of technology on modern perceptions of time, precision and speed, result in a disconnect from the essence of time as a metaphysical subject. In reality, we experience time as a variable. An hour on a Tuesday morning typically seems to go slower than an hour on Friday night. Behavioral psychologists Geoffrey Underwood and Rodney A. Swain observed as much in their 1973 article "Selectivity of Attention and the Perception of Duration" in the journal *Perception*. In it, they concluded that "those passages requiring more attention for analysis were judged to be of greater duration than those requiring less attention."[17] Nevertheless, technology enforces a mechanical constancy to time that defies our perception of it. The natural rhythms, cycles, and patterns of time, replaced by industrialized constructions, seem distant things on the modern landscape. The speed and precision at which the modern world moves is beyond human ability to keep up, and in many ways beyond human ability to comprehend. As Rebecca Solnit observes in *Wanderlust: A History of Walking*: "Modern life moves faster than the speed of thought. Our feet and our minds travel at 3 miles per hour."[18] The unintended consequences of this technological outpacing of human life are comprehensive. Aldo Leopold wrote in 1838 that "we end, I think, at what might be called the standard paradox of the twentieth century: our tools are better than we are, and grow better faster than we do. They suffice to crack the atom, to command the tides, but they do not suffice for the oldest task in human history: to live on a piece of land without spoiling it."[19] The modern preoccupation with precision that defies human comprehension and speed that defies human ability to keep pace as the defining characteristics of temporal understanding may come at the expense of any perspective that time itself may be running out.

TECHNOLOGY AND SPACE

In the story of *The Aleph*, Jorge Luis Borges described a piece of fictional technology that embedded in a particular time and place where one could view all times and places simultaneously. "In that single gigantic instant," Borges explained, "I saw millions of acts both delightful and awful; not one of them amazed me more than the face that all of them occupied the same point in space, without overlapping or transparency." He then described the Aleph and what he saw in it:

> I saw a small iridescent sphere of almost unbearable brilliance. At first I thought it was revolving; then I realized that this movement was an illusion created by the dizzying world it bounded. The Aleph's diameter was probably little more than an inch, but all space was there, actual and undiminished. Each thing (a mirror's face, let us say) was infinite things, since I distinctly saw it from every angle of the universe. I saw the teeming sea; I saw daybreak and nightfall; I saw the multitudes of

America; I saw a silvery cobweb in the center of a black pyramid; I saw a splintered labyrinth (it was London); I saw, close up, unending eyes watching themselves in me as a mirror; I saw all the mirrors on earth and none of them reflected me . . . I saw the Aleph from every point and every angle, and in the Aleph I saw the earth and in the earth the Aleph and in the Aleph the earth; I saw my own face and my own bowels; I saw your face; and I felt dizzy and wept, for my eyes had seen that secret and conjectured object whose name is common to all men but to which no man has looked upon—the unimaginable universe.[20]

Borges was writing fiction in 1945, but over the course of modernity, the *mythos* of the Aleph has in many ways become *techne*. The technological triangulation of space is increasingly comprehensive. From mapping territory, to mapping phenomena, to mapping ideology, technology has been instrumental in the triangulation of the modern world. Over the course of history, maps have served the dual purpose of being instruments of understanding and tools of control.

Mapping territory has been a human pursuit since prehistoric time. Cave paintings represent early landscapes and the ways in which humans interacted with them. In the ancient world, mapping was systematized by mathematical reasoning. Mesopotamian and Egyptian maps were both celestial and terrestrial, and the Greek geographer Ptolemy based his list of cities in the ancient world on the coordinates of latitude and longitude he reckoned by taking into account the calculation of the earth's circumference as previously calculated by Eratosthenes. Both the Roman Empire and the Han Dynasty in China applied the principle of centuriation for purposes of cadastral accounting, logistical planning, and bureaucratic subdivision. Maps for the Roman and Chinese emperors were tools of imperial control; the projected grid imposed economic, political, and martial control over the territory it bound.[21]

In the medieval world, maps became less literal and more symbolic, reflecting the shift toward a holistic religious worldview that blurred the lines between the material world and the spiritual world. Monsters appeared on maps where a lack of definite geographical knowledge led their makers to project their fears and prejudices of the world beyond their measure on to *Terrae Incognitae*. As with the Hereford World Map and other *mappamundi* from the Middle Ages, God is featured prominently at the top of the map, and Jerusalem is at the center, both the geographical and ideological axis upon which medieval Christendom spun. Cannibals and mythological beasts occupy the edges of the map, and biblical stories of the past share space with the impending last judgment of the book of Revelation.[22]

As maps tell us as much about the times in which they are made as they do the spatial information they convey, it follows that maps of the early modern era show a renewed commitment to technological development for the purposes of political and economic representations of space. The systematic surveying of space began anew and in earnest in fifteenth- and sixteenth-century Europe, in part due to a renewed interest in the science of astronomy and the principles of geometry as tools toward decoding a rational and universal order of things, and in other part due to an increasingly ontologically oriented power structure's preoccupation with using such information to reify and project their political and economic power onto space and the people who inhabit it. With the enclosures movement, cadastral mapping became a legal mandate. The formalized Ordnance Survey was established as an extension of military mapping in eighteenth-century England, in the wake of the Jacobite Rebellion in Scotland. In order to put down rebellion, territory must be put down on maps. The authority and propriety that cadastral mapping provided bound territory by "the law of the land," subjecting it and its inhabitants to imperial rule. The technological impetus of the Ordnance Survey was the three-foot Ramsden Theodolite, purchased by the crown for the Royal Surveyors "chiefly with a view to the more effectual execution of the work."[23] The technical authority of the survey was the basis for its legitimacy, and reinforced the authority of landowners and the crown through the completion of the Principal Triangulation of Great Britain, which was completed in 1853. Triangulations of territory carried on throughout the nineteenth century, as agents of empire bound the land by trigonometric surveys throughout Europe, North and South America, Africa, and Asia. *Terrae Incognitae* and the monsters that inhabited them gave way to precise maps that imposed political, economic, and martial authority over the territory they conveyed.

The technological capabilities of mapmakers shifted from mapping territory to mapping phenomena in the nineteenth century. Henry Drury Harness's 1837 map of Ireland superimposed statistical data onto the territory it represented, showing population density and traffic flow. Medical doctors began using maps to track the spread of disease. Thomas Shapter's 1849 map of cholera deaths in Exeter and John Snow's 1854 map of cholera deaths in London led to significant medical breakthroughs in understanding disease. Charles Booth's *Maps Descriptive of London Poverty* from 1889 and 1898–1899 led to radically new ideas regarding class and urbanization at the *fin de siècle*. These maps were essentially analog Geographical Information Systems (GIS). The advent and proliferation of computer technology in the latter part of the twentieth century saw a revolution in cartography away from static analog models toward more dynamic, interactive digital models. With GIS, anything can be mapped. Crime statistics, gas prices, school district test scores, median home incomes, weather patterns, the effects of radiation leaks, the migratory patterns of humans and animals, the effects of the tides on coastlines, the proliferation of community recycling, a particular sports team's fan base, or the spread of a specific disease. Moreover, this type of information can be layered, so that if a spatial correlation occurs between the two, say the effects of a radiation leak and the spread of a particular form of cancer, the pattern is easily identified in visual terms.

GIS have tremendous potential, but also have chilling implications. With the proliferation of the internet, GIS have shifted from mapping phenomena to mapping ideologies. The internet, theoretically conceived as "cyberspace" is mapped largely through the process of data mining. Marketers collect metadata to observe the habits and interests of consumers in the hopes of gaining a market advantage, while the state collects metadata to monitor threats to its sovereignty. In both cases, GIS map our ideas, our beliefs, our concerns, our fears and our hopes with pinpoint accuracy. Jonathan Harris's digital project "We Feel Fine" collects words connected to feeling and emotion on the internet, and projects them geographically. A user can compare how one particular place is "feeling" at any particular time, and parse that information by demographics like age and gender. Once locked in, the user can include weather patterns, news stories, and other information to determine the root cause of the way any particular place is "feeling" at any particular time. This is fascinating in its potential for human understanding, and terrifying for its potential for purposes of control and manipulation of both people and places.

While the Aleph of Jorge Luis Borges's imagination was something of a theoretical wonder when he wrote his story in 1945, the development of mapping techniques and systems over the course of modernity shows that Borges's literary *mythos* has increasingly become manifest in *techne*. It was a vision that Charles II had when he commissioned the Royal Greenwich Observatory in 1675, "so as to find out the so much-desired longitude of places for the perfection of navigation."[24] E. Walter Maunder considered the implications of this comprehensive triangulation of time and space at the turn of the twentieth century:

> So we have been led step by step from the mere desire to help the mariner to find his way across the trackless ocean, to the establishment of the secret law which rules the movements of every body of the universe, till at length we stand face to face with the mysteries of vast systems in the making, with the intimate structure of the stellar universe, with the apparent aimless, causeless wanderings of vast suns in lightning flight; with the problems we cannot solve, yet cannot cease from attempting, problems to which the only answer we can give is the confession of the magicians of Egypt—"This is the finger of God."[25]

NATIONALISM, URBANIZATION, AND TIME

The new ideas and perceptions of time that came with the proliferation of technology at the onset of the modern era were in many ways reinforced by the emergent concept of nationalism and the attendant force of urbanization. Like the advance of technology as a fundamental characteristic of modernity, both nationalism and urbanization changed the ways in which human beings inside of the project of modernity reckoned time. These new forms of reckoning obfuscated natural temporal rhythms and patterns which had evolved over the course of history, replacing them with new, hierarchically imposed models of temporal accounting that served to consolidate power and wealth among the few and disenfranchise the rest from the organic cycles of premodernity.

A nation, superficially defined, describes a group of people organized by common descent, language or history, organized as a separate political state, and occupying a defined territorial space. Underneath this veneer, however, is the idea that nations are inherently fictions; they are invented identities by which humans are ordered and arranged for the purpose of regulation and control. Benedict Anderson advanced this idea by suggesting that the nation is "an imagined political community."[26] Ernest Gellner argued that "there is nothing natural or universal about possessing a nationality."[27] To this point, the NASA astronaut William C. McCool observed from his orbital vantage point as commander of the Space Shuttle Columbia that "from our orbital vantage point, we observe an earth without borders, full of peace, beauty and magnificence, and we pray that humanity as a whole can imagine a borderless world as we see it and strive to live as one in peace."[28]

Nationalism is an inherently modern concept. Eric Hobsbawm wrote that "the basic characteristic of the modern nation and everything connected with it is modernity."[29] Josep Llobera went so far as to call the nation "The God of Modernity," arguing that the worship of the nation as an abstraction of political, economic, and cultural principles upon which people martyr themselves was something altogether religious in design. He explains: "It would be difficult to explain the appeal of modern nationalism without acknowledging the extent to which nationalist ideology is indebted to religion, albeit in modern times in secularized form."[30]

To promote the interests of nationalism, which are fundamentally bound up with the parallel conditions of modernity (those being technological advance and the proliferation of capitalist markets), its attendant bureaucracy of officials and agents construct an historical understanding of time that conforms to their agenda. These administrators, so many fifty-headed and hundred-handed Hekatonkheires, are principally responsible for what Eric Hobsbawm called "invented traditions, which reinforce the hierarchical model of modernity and root its legitimacy through "automatically imply(ing) continuity with the past." This historical past is constructed through the development of secular education, the invention of public tradition, and the mass production of public monuments, all of which establish continuity between a prenational past and a national present to legitimize the project.[31]

In reality, these constructed historical narratives isolate the national identity from any connection to larger historical conditions. The cracks in the façade of national identity were most readily apparent on the front in World War I, when soldiers in the terrible conditions of the trenches began to identify more with fellow soldiers on the other side of "no man's land" than with their commanding officers, who sent them over the top while staying safely in the trenches themselves. International class identity was more evident to the soldiers than class identity, when on Christmas Day, soldiers along the front emerged from the trenches to share in fellowship, song, bartering, and, by many accounts, a game of soccer. The military and political leadership on both sides downplayed and tried to altogether ban such fraternizing among soldiers from there on in.[32] Their goal was to break international class identity which would upend the eschatological vision being enacted on the battlefields, and replace it with a prejudicial hatred of a foreign "other" whose identity as an enemy and therefore a threat would keep people deferential and obedient to the emergent hierarchical order of modernity.

Benedict Anderson observed that the function of national histories is commemorated and reinforced "through mass media, the educational system, administrative regulations and so forth." The national biography is reified in museums, textbooks, and political programming to reinforce the invented tradition.[33] From this perspective, we can revisit the debate surrounding western civilization and world history as overarching historical narratives. The former, exclusive in nature, seeks to align Anglo-American modernity with the tradition of medieval European Christendom and root its primary political, cultural, and economic foundations in the "classical" histories of ancient Greece and Rome. This serves to fuel modern debates about what types of languages, races, ethnicities, and religions constitute a particular nation and therefore "belong" there. These are the elements of a phenomenon Hobsbawm referred to as "a consciousness of belonging."[34] The world history model is transnational, and by definition inclusive. Everyone belongs and contributes to the historical narrative. The western civilization model is national, and by definition exclusive. Certain people belong, others do not. From this come debates about "illegal immigrants," English language requirements, which prayers are acceptable in public and which are not, what is acceptable dress, and so forth. It is firmly entrenched in the concept of national identity, and a model of historical time that is ultimately invented and constructed to suit the interests of the modern power structures which it serves.

Urbanization as the second factor in the organization of human populations has also affected perceptions of time over the course of modern history. As a historical force, modern urbanization is a direct result of the Industrial Revolution, a broadly conceived shift in the history of work and labor which most historians agree began in the English midlands in the middle of the eighteenth century. As the effects of enclosures took hold, deracinated populations moved into newly formed industrial cities like Birmingham, Leeds, Manchester, and Sheffield, in which the bourgeoisie had begun to build mechanized factories due to their proximity to power supplies, particularly coal.[35] With the growing labor force that came into cities as a result of the dual forces of enclosures and industrialization, the bureaucracy of management expanded to organize workers and enforce the terms and conditions of their employment. Central to the management of the industrial work force was the clock.

To coordinate industrial workers on clock time required an organized effort to break the habits and patterns of agricultural work, which were variable and rhythmic, and replace them with a mechanical constancy governed by "king minute." The enforcement of "time discipline," and the proliferation of clocks and watches among the laboring classes occurred, as E. P. Thompson suggests, "at the exact moment when the industrial revolution demanded a greater synchronization of labor."[36] Public clocks were ubiquitous by the end of the sixteenth century, as modernity's attendant systems began their exponential rise, but were largely attached to and maintained by churches. Jacques Le Goff's attention to the shift from eschatological and spiritual "church time" to cyclical and material and commodified merchant time is informative in this context, as in it lie the roots of the moral ordering of time toward ends that serve the emerging industrial order.[37] Church bells became factory whistles, and the shift from the temporal order of medieval towns to early modern industrial cities was complete. Most of the factory whistles of the first wave of industrial modernization have since gone silent, replaced by alarm clocks, watches, and cellular phones which offer the time as an ever-present reminder that time discipline is central to the ontological stasis of the modern condition. It is a pattern with roots in urban industrial constructions of time, inherently artificial and imposed for the purposes of hierarchical ordering and control.

The dual forces of nationalism and urbanization, together representing a fundamental shift in the social, political, and economic organization of human beings, have had a profound effect on the ways in which we construct our understandings of time in the modern world. From the construction of histories that legitimize the hierarchical ordering of people within a system of nationalist inclusion and exclusion, to the enforcement of time discipline within the framework of growing cities organized around the emergent technologies of mechanized industrial production, time is a fundamentally different thing in the modern world than it was in the ancient and medieval worlds. Separated from our universal past as inhabitants of the planet, and detached from the natural rhythms of work, play, and life that we seek to revisit in various forms of "recreation" and "vacation," our alienation serves the interests of ontological hierarchical order at the expense of our health, our well-being, our sanity, and our very lives. It is the reason we travel to distant places to remind ourselves of our common humanity, leave our watches on the nightstand and take pictures of sunsets after we have left the office. These are only temporary escapes from the spatial and temporal confines of the technocratic monoculture, however, sanctioned through work policies and official passports, to provide incentive to work even harder for the next year to again be able to vacate the our obligations and live on our own time, even if for just another week or two.

NATIONALISM, URBANIZATION, AND SPACE

Like time, the dual forces of nationalism and urbanization have had a profound impact on modern spatial conditions. With the delineation of national boundaries came an increase in the precision, detail, and importance of the map as a defining characteristic of national identity. Here, the confluence of technology, nationalism, and market economies comes together to define a specifically national space within which hierarchically authorized transactions and interactions are monitored, regulated, and controlled. With the onset of modern urbanization, a shift from small, freehold agricultural production to large-scale, wage-based industrial production paralleled a shift in modern identity from domestic production to domestic consumption. Together, nationalism and urbanization as dual forces transformed the manner in which modern populations are organized and have fundamentally changed the social, political, and economic conditions in which people regularly interact.

To be certain, the concept of the boundary is not exclusively a product of modernity. It has long been the preoccupation of rulers to survey their territories and triangulate them for the purposes of power and control. Some of the earliest written documents discovered of the Sumerian civilization in Mesopotamia are boundary markers, or histories recording the clash over boundary disputes. This is the case of the Sumerian inscription of Umma and Lagash, which dates from about 2500 BC. It sets up the contest between the two cities by recounting a border clash:

> By the immutable word of Enlil, king of the lands, father of the gods, Ningirsu and Shara set a boundary to their lands. Mesilim, King of Kish, at the command of his deity Kadi, set up a stele [a boundary marker] in the plantation of that field. Ush, ruler of Umma, formed a plan to seize it. That stele he broke in pieces, into the plain of Lagash he advanced. Ningirsu, the hero of Enlil, by his just command, made war upon Umma.

After a description of the battle that followed, the stele concludes:

> Ili, took the ruler of Umma into his hand. He drained the boundary canal of Ningirsu, a great protecting structure of Ningirsu, unto the bank of the Tigris above from the banks of Girsu. . . . he commanded that the boundary canal of Ningirsu; the boundary canal of Nina be ruined. . . . Enlil and Ninkhursag did not permit [this to happen]. Entemena, ruler of Lagash, whose name was spoken by Ningirsu, restored their canal to its place . . ."[38]

With this, we see that the concept of boundaries and disputes over their placement are not exclusively modern phenomena. What separates the modern epoch from the ancient and the medieval are the systematic formalization of national boundaries which developed parallel to technologies and the expansion of territorial empires in the sixteenth century and through the nineteenth century. As we have already noted, medieval maps were an amalgam of religious and geographical information; formal maps of medieval kingdoms simply did not exist. Kingdoms were essentially a succession of places where a ruler could assert his power, an idea Michael Biggs defines as "the dynastic realm."[39] The dynastic realm was fluid and porous. Borderlines did not exist so much as borderlands did. The conceptual apparatus for mapping as a tool for projecting authority and control over territory, along with the technical capability for doing so precisely and effectively, is a product of Renaissance thought. The application of the delineated modern map as a means of imposing territorial hegemony is rooted in the martial and economic competition of the emergent imperial powers of Western Europe during the same time.

The modern map emerged as a projection of national identity at the same time it was reified by technological advances and sponsored by the state for purposes of economic authority and control. Surveyors sent by the British, French, and American governments competed to acquire territorial information through the process of "exploration" through the seventeenth, eighteenth, and nineteenth centuries, in an effort to legitimize their claims through the triangulation and integration of space into the imperial and ultimately the national identity. Through the Cassini Survey in France, the Ordnance Survey in Great Britain, and the Land Ordinance of 1785 in the United States, the systematic accounting and incorporating of space turned the map into an avatar for the nation and national identity. The map which Baudrillard observed as preceding the territory came to represent it. In the process, it reinforces not only the idea of nationalism, but also the technological and economic systems which are at the origins of its design. At the confluence of these systems is the condition of modernity, framed within precisely drawn borderlines, to impose power and control over space and the people who inhabit it. To this point, Ambrose Bierce defined the term "boundary" in his 1911 satirical reference book, *The Devil's Dictionary*:

> Boundary, *n*. In political geography, an imaginary line between two nations, separating the imaginary rights of one from the imaginary rights of others.[40]

Like nationalism, urban spaces have the capability of rendering people subject to the terms and conditions of modernity. The shift from freehold or manorial agricultural production that came with the process of enclosures in the fifteenth and sixteenth centuries deracinated large cross sections of the population, many of whom eventually relocated to cities at the onset of the Industrial Revolution. The domestic production of food, clothing, and shelter shifted to industrial production of commercial goods, for which workers received wages to buy those things that they used to provide themselves. As Karl Marx explained the process in *Das Kapital*, "whilst the place of the independent yeoman was taken by tenants at will, small farmers on yearly leases, a servile rabble dependent on the pleasure of the landlords, the systematic robbery of the Communal lands helped especially, next to the theft of State domains, to swell those large farms, that were called in the eighteenth century capital farms or merchant farms, and to 'set free' the agricultural populations as proletarians for manufacturing industry."[41]

Thus, domestic producers became commercial producers and, by proxy, domestic consumers with the onset of modern urbanization. Self-sufficiency and barter was replaced by the industrial wage, which the newly urbanized proletariat could take to the urban market to purchase things which they had previously been able to produce and procure domestically. The urban market is both an ideological and physical space. Early modern cities, like their medieval predecessors, had commercial spaces, market places, squares, and streets woven into the fabric of residential districts, where the surplus of industrial production was transacted. Over time, those commercial districts became increasingly isolated from residential neighborhoods, monitored and controlled to promote access to a certain cross section of the population, and to prevent access from others. The promotion of "downtowns" as commercial centers allowed for the privatization of commercial space, along with its constant monitoring and regulation.[42] Zoning laws, which force the centralization of consumer activity, further disenfranchised the proletariat from access to the goods and services once held in common and for which they could provide of their own accord. Centralizing wealth and commerce allowed for the marginalization of poverty in urban spaces. The slums, ghettoes, and favelas of modern cities isolate the poor geographically, compounding their social lack of access to capital through physically distancing them from it.[43]

Attempts to reinvent urban spaces as centers for domestic production rather than consumption have been difficult to establish, as self-sufficiency ultimately challenges capitalist hegemony. Urban farming is illegal in many cities, but in those blighted by the demise of industry in the late twentieth century, it is seen by many as a path to renewal and revitalization. The City of Detroit is experimenting with urban farming in the wake of its corporate economic collapse, but in most large cities, urban agriculture is problematic, as it challenges the fundamental cycle of consumer dependence that capitalist markets depend on.[44] Similarly, cities ravaged by the collapse of the corporate industrial economy at the turn of the twentieth century are taking tentative steps to reestablish forms of commoning as a way to repopulate its urban centers and maintain what is left of the metropolitan infrastructure. By making unoccupied homes available for as little as one dollar in some cases, city governments are taking tentative steps to bring people back to the hollowed out cities and rebuild what the industrial and banking economies destroyed.[45]

Taken together, the modern forms of national and urban space have had a profound effect on the human condition. National spaces are rooted in the advance of technology and the pursuit of imperial territory and wealth. Their fixed borderlines betray the regional identities that flow across them, imposing instead a fixed system of political, economic, and cultural conditions through official regulation, legal administration, and martial enforcement. Urban spaces are designed to enforce the mode of commercial production and consumption, to which many people within them have no access due to the disparity of wealth that market capitalism inherently produces.

MARKET ECONOMIES AND TIME

Over the course of modern history, the market economy has fully integrated itself into nearly all aspects of life. Perhaps nowhere is this more obvious than the ways in which the market economy has come to affect and alter the perception of time. The degree to which markets have shaped our understanding of time, and how the agents of the market economic system have organized temporal dynamics to accommodate its proliferation, is evident in the historical consideration of the modern timescape. Market economic transactions, once relegated to specific times, proliferate over the course of modern history to become a predominant feature in how human beings reckon the temporal condition.

There was a fundamental distrust of market economics as a practical model of exchange in the ancient world. Hesiod warned in *Works and Days* that "money should not be seized; that gold which is God's gift is better. If a man gets wealth by force of hands or through his lying tongue, as often happens when greed clouds his mind and shame is pushed aside by shamelessness, then the gods blot him out and blast his house and soon his wealth deserts him." He drove his point home a few lines later: "Shun evil profit, for dishonest gain is the same as failure."[46] In Book One of *Politics*, Aristotle wrote that the idea of retail trade as a form of money-making was "a kind of exchange which is justly censured; for it is unnatural, and a mode by which men gain from one another."[47]

Ancient markets were not only spatially designated, as we shall soon see, but temporally designated as well. The Athenian markets, the *emporia*, were open for business only in the morning hours. T. G. Tucker explained that "what we should call ten o'clock was called by the Athenians 'full market.' About noon the stalls and wickerwork booths are cleared away and the ordinary business part of the day is done."[48] Markets were regulated by several levels of official control. *Metronomoi* set the weights and measures, *sitophylakes* regulated prices, *agoranomoi* monitored the purity of the products in the market.[49]

The Roman *nundinae* was a day specifically designated for market economic activity.[50] While the centers of commercial activity were open daily for trade, they were busiest on the *nundinae* which occurred every ninth day. Macrobius recounts in the first book of *Saturnalia* that official efforts were made "with such wary precision" to separate market day from both religious and secular holidays.[51] In addition, neighboring towns staggered their market days to avoid conflict and competition. Ancient markets were temporally regulated by both law and custom.

As Jacques Le Goff suggests, the Catholic Church viewed medieval merchants suspiciously because "their profit implied a mortgage on time, which was supposed to belong to God alone."[52] Nevertheless, feudal lords began aggressively acquiring charters for village markets beginning in the thirteenth century. Charters regulated the conditions of markets, including the times they could be held. In 1282, King Edward I granted Henry de Lacy, Earl of Lincoln, a charter to hold Saturday markets in the Borough of Congleton. He also granted William de Tabbeleye a charter for weekly Saturday markets in Knutsford ten years later. In 1241, Henry, Duke of Lancaster held a charter sanctioning markets on *nundinae*, clear evidence that the medieval concept of market regulation was rooted in Roman custom and law.[53]

The establishment of "blue laws" at the time capital was expanding globally in the early modern era is significant. Their history is linked to religious concerns that commerce should not interfere with worship on holy days. The idea is based on the fifth biblical commandment "Six days shalt thou labour, and do all thy work. But the seventh day [is] the sabbath of the LORD thy God: [in it] thou shalt not do any work, thou, nor thy son, nor thy daughter, thy manservant, nor thy maidservant, nor thy cattle, nor thy stranger that [is] within thy gates. For in six days the Lord made heaven and earth, the sea and all that in them is, and rested the seventh day; wherefore the Lord blessed the sabbath day, and hallowed it."[54] The emperor Constantine enforced the Sabbath as a day of rest from labor in AD 321 with this decree:

> All judges and city people and the craftsmen shall rest upon the venerable day of the sun. Country people, however, may freely attend to the cultivation of the fields, because it frequently happens that no other days are better adapted for planting the grain in the furrows or the vines in trenches. So that the advantage given by heavenly providence may not for the occasion of a short time perish.[55]

This decree against work and commerce on Sundays was reinforced by the Anglo Saxon king Ina in 691, Alfred in 900, Athelsane in 924, and by the English kings Henry III in 1237, Henry VI in 1444, James I in 1606, and Charles II in 1676.[56] Virginia was the first to pass Sunday legislation in the English North American colonies in 1610. A first offense was punished with a fine, a second time offender could be fined "and also be whipt," and a third time offender could "suffer death."[57] Similar laws were passed throughout the colonies, and the market economy was legally held in check until the rise of corporate capitalism in the nineteenth century, when corporations began challenging the legality of "blue laws" in the courts, and most state repealed their blue laws in the latter part of the twentieth century to make way for the expansion of the market economy into every day of the week. While a patchwork of blue laws still exists in local communities, mostly related to the sale of alcohol, the market economy is an aspect of modern life on all seven days of the week.

Increasingly, it is also an aspect of life during all twenty-four hours on each of those days. Twenty-four-hour stores were mostly an indirect result of factory shift work as the consumer economy expanded during the latter part of the Industrial Revolution. The shift to 24/7 market access has most obviously been aided by the expansion of the internet. Markets never sleep. In 1985, regular stock trading hours in the United States were from 9:30am to 4:00pm Eastern Standard Time. In 1998, with the rise of online investing, "Pre-Market" trading hours were added, from 4:00am to 9:30am. In addition, "After-Market" hours were added from 4:00pm to 8:00pm.[58] After 8:00pm EST, investors can turn their attention from the American Markets to the Japanese Markets, which open their second session of trading at 7:30pm EST. Economic analyst Douglas McIntyre, who writes for an online news site called *24/7 Wall Street*, concludes that "the reasons for U.S. equities to be traded on a 24-hour cycle are countless. What does it matter if a few people need to work in the dark?"[59]

While the temporal cycle of the market is continuous and self-perpetuating in the context of global capitalism, it is also divisible to the nanosecond. Increasingly fast computer technology makes the market trading technique known as nanotrading possible. The corporation Fixnetix developed a microchip in 2011 that processes trades in 740 nanoseconds. While the technical accomplishment is impressive, the result is a new market phenomenon called "flash crashes." The first occurred on May 6, 2010, when the Dow Jones stock index fell 600 points in six minutes, between 2:42 and 2:48 EST. Between 2006 and 2011, 18,520 crashes and spikes had occurred below the 950-millisecond level. The physicist Doyne Farmer, who works in the fields of chaos theory, complexity, and econophysics, points out the obvious: "It's hard to think these things through, because nobody understands them."[60]

Even the words we use to speak about the concept of time are economic. We *spend* time, we *save* time, we *waste* time. Seldom do we *take* time, often because we cannot *afford* to do so. Market economics have become an increasingly pervasive factor in the ways in which we perceive and conceive time within the context of modernity, and this has had a noticeable effect on the effect of time as a historical phenomenon. Time is increasingly economized, and we are left to reckon its costs. E. P. Thompson quoted Stephen Duck's 1736 poem "The Thresher's Labour" at length in his important article on the evolution of time during the Industrial Revolution, but one verse in particular, which Thompson edited for brevity, deserves repeating in full, with attention to the overarching thesis of this work and its allegorical foundation in the Titanomachy of Hesiod:

> Now in the Air our knotty Weapons fly,
> And now with equal Force descend from high;
> Down one, one up, so well they keep the Time,
> The CYCLOPS' Hammers could not truer chime;
> Nor with more heavy Strokes could *Aetna* groan,
> When VULCAN forg'd the Arms for THETIS' Son.
> In briny Streams our Sweat descends apace,
> Drops from our Locks, or trickles down our Face.
> No Intermission in our Work we know;
> The noisy Threshal must for ever go.[61]

MARKET ECONOMIES AND SPACE

As market economies have become more temporally pervasive over the course of modernity, so too have they expanded into space. Markets have moved from the bounded and peripheral geographical place of their historical conditions to occupy a central location in modern life. They have become ubiquitous in the process, expanding not only geographically, but ideologically as well, becoming central to both the theory and practice of modern condition. As Henri Lefebvre observed in his seminal work *The Production of Space*, capitalist space functions to promote "the hegemony of one class." The authority of the marketplace as the preeminent spatial discourse in the public sphere is the connection of power and place, where attempts to undo, challenge, or otherwise resist the hegemony of the marketplace is inherently subversive.[62]

Again, the ancient and medieval worlds inform our understanding of the modern. Ancient markets were not only relegated to specific times, but also specific places. The economist Karl Polanyi argued that international ports of trade dating back to the second millennium BC were intentionally peripheral to towns and cities. Their separation, Polanyi asserted, was because trade was mediated, "while competition was avoided as a mode of transaction. Where present, it was relegated to the background, or was merely lurking on the periphery." Like the physical places where trade occurred, competition was marginal. The term "emporium" according to Polanyi, "conveys a meeting place of traders, located outside of the gates of a town, or even on an uninhabited coast."[63] Polanyi accounts for this in archaeological and anthropological terms in Mesopotamia, Asia Minor and the Black Sea, and later in the African kingdoms of Whydah and Dahomey, the Aztec-Maya empires of Central America, and in places throughout India and China, along with inland caravan cities like Palmyra and Kandahar.

The Greek model of the emporia takes its meaning from these origins, and was something altogether separate from the rest of the city. The Athenian Agora and similar emporia throughout the ancient Greek city-states were delineated by boundary stones which constrained commercial activity to within their spatial limits.[64] Eventually, the activity inside of the Agora became unbound, and political activity fused with the commercial to make it a significant, even central, public space in ancient Athens. This fusion of the political and the commercial, at a time when Athens was becoming the ascendant commercial empire of the Aegean world, seems consistent with LeFebvre's view that "society as a whole continues in subjection to political practice—that is, to state power."[65]

The revival of the medieval town in the eleventh and twelfth centuries was largely a result in the increase of trade that came as transportation and communication ties began to evolve on a regional and, with the Crusades, even international basis. "Market Towns" began to have defined roles in regional economies, and as we have seen, were sanctioned and regulated by royal decree for exclusive benefit. R. H. Hilton observed that "in all there were about twenty-five [county towns of Worcester and Warwick], the majority of which were founded or sponsored by lay or ecclesiastical lords in the thirteenth century. They were clearly a response to a combination of factors operating in the rural economy, an improvement in productivity, a considerable increase in population, an increased demand for manufactured commodities by agricultural producers and, very important, an increasing demand for rent and tax in money by lords and the state."[66] In this context, markets became both more spatially and conceptually central to the political, economic, and cultural conditions of the times and places in which they developed. Transient fairs were replaced by more permanent market squares, and the once-marginal commercial activity became vital to the functioning of the town through the twelfth and thirteenth centuries.[67] The movement of the place of the market economy from the periphery to the center, in both physical and ideological terms, is bound up with the ascension of Lefebvre's "hegemony of one class." The same lords who controlled the economic direction of this evolving world controlled the political infrastructure which sanctioned and officiated it, and by proxy, the cultural terms and conditions in which the economic and political systems manifest.

Modern industrial cities inherited and institutionalized the centrality of the market in physical and ideological terms. Like their ancient and medieval antecedents, the markets of modern cities were bound and defined in spatial terms. As Robert M. Fogelson explains: "In virtually every city downtown had some sort of physical boundaries, usually a bay, a lake, a river, or, in a few cases, a combination of them. But nowhere did these boundaries define downtown with precision." The elements that defined downtown were its centrality and dimensionality. "Although hard to define, downtown was easy to locate. It was the destination of the street railways . . . elevated railways . . . railroad terminals. Downtown was the home of the tall office buildings . . . department stores . . . downtown was the only part of the city wired for electricity."[68] Downtowns, in short, were designed to be centers of economic activity.

As quickly as modern cities rose, the merchants that built them evacuated. The confluence of racial and class identity made for volatile urban settings which threatened the security of the hegemonic class. The result was suburbanization, made possible by the technology and infrastructure of automobility. Suburbanization, in turn, gave rise to the supermarket, the strip mall, and the branch office, in other words, the physical proliferation of the marketplace. With the expansion of domestic markets came the expansion of international markets. The age of automobility was also the age of economic globalization. The Bretton Woods Accord, which established the International Monetary Fund (IMF), was signed in 1944 to ensure the sanctity of market economics as the primary form of social organization in a postwar environment. The result was the establishment of a unitary global market, governed by the IMF and the World Bank, to ensure that every place was subject to the terms and conditions of marketplace. The ubiquitous centrality of the market, and the criminalization of anything that challenges it, is a phenomenon Henri Lefebvre considers in his analysis of *scene* and *obscene*. He elaborates: ". . . walls, enclosures and facades serve to define both a *scene* (where something takes place) and an *obscene* area to which everything that cannot or may not happen on the scene is relegated: whatever is inadmissable be it malefic or forbidden, thus has its own hidden space on the near or far side of a frontier."[69]

As markets proliferated physically in the age of automobility and globalization, the digital technology that followed allowed markets to expand comprehensively, from the physical to the ideological realm. The privatization of physical space relegated the manifestation of resistance to the marginal periphery, advertisements that promote the scene of corporate-sanctioned consumerism are legal and acceptable, whereas graffiti or political protests that offer a critical challenge to the hegemony of the scene are criminalized and unacceptable: obscene. They are not permitted to remain in place for any amount of time, as they obstruct the fundamental purpose of modern timespace. The pervasiveness of digital markets, and the ability of its architects to extract metadata from their subjects, renders consumers servile in a sort of Slavery 2.0, where what Christian Parenti calls our "informational doppelgangers" are bought and sold by overseers without any recourse on the part of the individual whose thoughts the metadata comprises.[70] Similarly, our digital metadata tracks not just our comings and goings in physical space, but our ideological comings and goings: our feelings, beliefs, ideas, opinions, fears and hopes, which are all for sale on the capitalist market.

Attempts to resist the scene of physical and ideological market omnipresence/omniscience are met with the same fervor that early modern slave revolts were. The actions of Julian Assange, Chelsea Manning, and Edward Snowden, modern day Nat Turners, who sought to resist the slaveocracy by subjecting their overseers to the same means of control that they inflicted on their subjects, were roundly met by Olympian calls for summary execution and death. It is acceptable and sanctioned for the Pantheon to observe the scene from high Mount Olympus, but to suggest that Olympus be subject to view from below is treasonous, as it suggests the institutionalization of the obscene, which is meant to be kept behind the walls of Tartarus, guarded and kept out of view by the Cyclopes and Hekatonkheires.

From the marginal periphery to its ubiquitous physical and ideological presence, the market economy has had a significant effect on the conditions of space in the context of modernity. As with time, markets have slipped from their restrictive spatial bonds to become a central feature which defines the way humans interact. Markets, by their very nature, are predicated on inequality. Means and access are not universal. Those not included within the scene are relegated to the obscene. Like the commodities that are transacted within them, space too is commodified; it is produced and consumed. This economic operation is inherently exclusive and limited. As LeFebvre concludes: "Today everything that derives from history or from historical time must undergo a test." That test, within the context of this consideration, is the ways in which democratic ideology challenge the exponential manifestations of technology, nationalism and market economies within the temporal and spatial realms.

DEMOCRATIC IDEOLOGY AND TIME

As these forces of modernity, technology, nationalism and urbanization, and market economics, have appropriated the temporal realm for the purpose of control, the presence of a democratic ideology has served to check their advance and protect the temporal realm from complete seizure. At the vanguard of the fight has been the labor movement, to which time has been an integral part of its platform since the onset of industrialization. As the labor movement has conceded temporal issues, its fortunes have fallen, and its successes have been less numerous. Nevertheless, without the efforts of the working classes to protect the universal resource of time from privatization and hegemonic control, the modern world would resemble the dystopian hellscape of the most ambitious science fiction writers.

Working people have a long history of resistance to the control of time by the propertied classes. Perhaps one of the most interesting is the institution of Saint Monday. With the institution of the work week as a set temporal cycle at the onset of the Industrial Revolution, workers who previously set their own schedule in a craft-based, task-oriented economy rebelled against the regularization of work by simply not turning up for work on Mondays. As time discipline became more pervasive in industrial capitalism, workers would show up at their places of employment, but do little to no work on Mondays. Douglas Reid observed in his seminal study of the institution: "Saint Monday posed a cultural problem which went to the heart of the ruling order. . . . Since the application of power represented progress and moral improvement was also progress, amusements which were an adjunct to both were especially desirable to reformers whose 'great object' was 'the educational, moral and political improvement of the people.'" By the latter part of the nineteenth century, as industrial organization imposed its attendant modes of surveillance and control on working people to render them subject to the factory machines and the profit motives of their employers in the name of "progress" and "improvement," Saint Monday withered. As Reid concluded, "only its ghost lingered on."[71]

With Saint Monday martyred at the altar of industrial progress, the labor movement conceded the authority of the clock and turned its attention to time management. The eight-hour workday movement which came to unify the international working class in the late nineteenth and early twentieth century predicated its goals on a separation of work from life. In 1834, the English mill owner Robert Owen developed the tripartite division of the day based on clock time to head off the syndicalist movement at his factory. He founded a new society of employers and employees, the National Regeneration Society, to champion the cause of "Eight hours labor, eight hours recreation, eight hours rest."[72] Over the course of the nineteenth century, the international labor community had taken up the cause of the eight-hour day in reaction to the increased demands of industrialization on workers' time. Establishment of the eight-hour day was incremental throughout the early nineteenth century, specific to different trades in different countries. Article 123 of the 1917 Mexican Constitution established the eight-hour day and a living wage for Mexican citizens. Both Portugal and Spain established an eight-hour day in 1919, as did France in 1936. In the United States, the Fair Labor Standards Act passed in 1937 established the eight-hour day as a standard for industrial work.[73]

The shift to a postindustrial economy has led to the erosion of the eight-hour workday. In the United States, many workers clock longer hours to maintain their benefits package in a hypercompetitive environment, while many other workers are relegated to "part-time" employment so their employers do not have to pay them benefits. These workers often work two or three part-time jobs in exchange for wages that do not keep up with the cost of living. In the meantime, worker productivity has increased exponentially. The economist Erik Rauch asks at the start of his 2000 essay *Productivity and the Workweek*: "What if, instead of using productivity increases to buy more possessions, we used them to get more time instead?" Rauch asserts that "even since 1975, supposedly an era of low productivity growth and stagnation in living standards, officially measured productivity has increased almost 70%. The average worker would therefore need to work only 23 hours per week to produce as much as one working as recently as 1975."[74] This has led many economists to call for legislation shortening the workweek and increasing the wages of workers to a "living wage," while mandating benefits and pension packages for all workers. According to Anna Coote, Andrew Sims, and Jane Franklin, who authored the New Economics Foundation report *21 Hours*, "A 'normal' working week of 21 hours could help to address a range of urgent, interlinked problems: overwork, unemployment, over-consumption, high carbon emissions, low well-being, entrenched inequalities, and the lack of time to live sustainably, to care for each other, and simply to enjoy life."[75]

The original cause of the eight-hour workday drew concerns from employers and the propertied classes that workers would "abuse the privilege of more leisure." Workers having "free time" were a fundamental threat to the institutional leisure class, so controlling the time of workers when they were not working, through the institutionalization of work ethic and its attendant systems of morality became the goal to render workers submissive and obedient to the will of the employers. The institutionalization of education became the perfect training ground for future workers, and it is in the schools that workers/consumers are trained, under the rhetorical guise of educating citizens. Children are forced to sit still, obey bells and alarms, ask permission to physically relieve themselves, and increasingly take more and more work home to train them for "life in the new economy." Recess, or "free time," is minimized, leaving children listless. This continues in spite of every study that affirms the understanding that children ought to, as Julia Bishop and Mavis Curtis put it, "satisfy their psychological and physical needs and develop their own social networks; social, cognitive and artistic skills; and imagination and creativity."[76] Those children who cannot restrain their energies are often diagnosed with ADHD, and pharmaceutical corporations provide drugs to render them passive and obedient, or in the words of the corporate-educational complex, "a good student."

Over time, this institutional training has had its effect. Workers identities shifted from active citizens engaged in a struggle for social justice to that of passive consumers. As Michael Kazin and Steven J. Ross explain, even the American holiday of Labor Day shows a shift from active militancy to passive acceptance. Labor Day was institutionalized in 1882 unite the wide variety of working class identities in the United States, not all of whom identified with the militant anticapitalist stance of the international May Day celebrations.

Nonetheless, early Labor Day celebrations were infused with class antagonisms. As Ross and Kazin explain, however, is that the celebration has, over time, become "politically anesthetized," a product of the sublimation of working class identity by that of consumerism.[77]

And so, these battlegrounds of time, the school and the home and the workplace, become the places where the reclamation of teleological progress begin anew. Increasingly, parents are insisting that their children be educated to be active and engaged citizens, not trained to be passive and apathetic consumers, and that "free time" be an integral part of the school day.[78] At home, new ideas like the "slow food movement," a reaction to a fast food culture which sees Americans eating 20% of their food in their cars and often eating at their desk while they work, are taking time back for the things that people, rather than the abstract economic indicators, actually value.[79] The workplace, however, is still a place where temporal liberation is difficult to achieve. The United States is the only "advanced" economy that does not require employers to provide paid vacation time, and maternity leave is often unpaid if available at all. Other nations are much more ahead of the United States in both regards. In the United Kingdom for example, workers are legally entitled to 5.6 weeks paid holiday per year. Statutory maternity leave is fifty-two weeks, with thirty-nine of those weeks being eligible for some amount of pay.[80] Despite most people's insistence that they would like to work less and earn more, the attendant power structures of modernity align to prevent this democratic vision from becoming a reality. Woody Guthrie's lyrics to the song "Talking Union," written in 1941 when his group, The Almanac Singers, was helping to organize the Congress of Industrial Organizations unions, are reminiscent of the long struggle for temporal autonomy. The song promises temporal liberation in the form of, among other things, shorter hours and vacations with pay as benefits of union membership. The last lyric of the song is a rallying cry in the battle for temporal liberation in this modern age: "Take it easy, but take it."

DEMOCRATIC IDEOLOGY AND SPACE

As democratic ideology can challenge the ontological temporal stasis imposed from above, so too can it defy the triangulation of space which maintains control in the interests of hierarchical order. The first manner of reclaiming space is by asserting a present, overt, embodied resistance to the technocratic monoculture and its attendant enforcement mechanisms. The second way of reclaiming space is through the assertion of obfuscated, hidden, abstract, and cryptic resistance. Both of these tactics are present and developed in the struggle for teleological resistance. The former goes by the name Occupy, the latter is known as Anonymous. While both movements are twenty-first-century creations, they have long roots in the history of struggle against the hegemonic powers that contribute to their efficacy. On September 17, 2011, on the ten-year anniversary of the reopening of the stock markets after the September 11th attacks on the United States of America, an estimated 1,000 protestors collected on Wall Street, and that night, about 100 of them slept in Zucotti Park, a "Publicly Owned Private Space" in Lower Manhattan, formerly named Liberty Park Plaza. They protested the economic inequality that the markets had exacerbated since the financial collapse of 2008, and from this the phrase, "We Are The 99%" came to signify the many who suffered as a result of economic globalization for the benefit of the few, or the 1%. Within a few weeks, Occupy protests had spread to over 600 locations in eighty-two countries, involving millions of people on every populated continent on Earth.

The army of Kronos had pushed itself out of Tartarus, past the Cyclopes and Hekatonkheires, and camped at the foot of Mount Olympus itself. The references were overt. Healthcare activist Sylvia Moore wrote about the injustices of the health care industry in October 2011: "A small group of billionaires operating in shiny glass towers with virtually no accountability to the public are deciding who lives and who dies. These health insurance executives—denizens of Wall Street—have set themselves up as Greek gods on Mount Olympus. It's time to topple them from that perch, and turn our healthcare system over to the people."[81] A play inspired by the movement, titled *Occupy Olympus* opened at the Fringe Festival in 2013. Based on Aristophenes's play *Plutus, God of Wealth*, which called into question the distribution of wealth in ancient Greece, the playbill for *Occupy Olympus* promised to "take on Zeus and the Olympian 1%."[82]

The Occupy protests were pushed back from Olympus with force. A coordinated effort, orchestrated by mayors of eighteen cities across the United States, cleared the camps with forced evictions and arrests in November 2011. The Occupy Camp at Zucotti Park was cleared as part of the crackdown by the New York Police Department working at the behest of Brookfield Office Properties, which owned the "Publically Owned Private Space" that was Zucotti Park, citing "health and safety reasons."[83] Similar actions occurring on a global scale closed the last remaining sites by February 2012.

The Occupy Movement, which began in 2011, has international roots in the Spanish Indignados movement, which was in turn inspired by the events of the Arab Spring in 2010 which saw millions take to the streets in Algeria, Bahrain, Djibouti, Egypt, Iraq, Jordan, Kuwait, Libya, Mauritania, Morocco, Oman, Palestine, Saudi Arabia, Sudan, Syria, Tunisia, Western Sahara, and Yemen. But the history of public protest has much deeper roots than these, and is a persistent theme in the struggle against hegemonic power. In ancient Rome, plebeians took to the streets in a form of protest known as *Secessio Plebis*, where they would abandon their work and shut down the city's commerce. Their numbers made this both a difficult strategy to organize and an effective tool of redress when they were able to unite and execute the protest. History accounts for five *Secessio Plebii* in Roman history, from 494 through 287 BC. These revolts increased the power of the plebeian classes within the Roman political, economic, and social orders, ultimately resulting in the *Lex Hortensia*, which gave resolutions passed by Plebeian Council the force of law.[84]

The American Revolution began in the streets of Boston, and the French Revolution began in the streets of Paris. These were all initially occupations of public space by people seeking redress from power in one form or another. The Trade Unions Movement, The Women's Suffrage Movement, and the Civil Rights Movement have all depended on a physical presence in public space to achieve their goals in the course of modern history. As the architect

Le Corbusier wrote in his 1948 exposition on the anthropometric scale of proportions, *The Modulor:* "To take possession of space is the first gesture of living, men and beasts, plants and clouds, the fundamental manifestation of equilibrium and permanence. The first proof of existence is to occupy space."[85]

Another means of resistance is secrecy and anonymity. Where the Occupy Movement depends on the assertion of a mass of bodies in a public space, Anonymous is an international network of activists who work in cyberspace. Anonymous targets the power structure online, challenging its presence there by subverting its ideology through information leaks, hacks, and other forms of digital attacks. Their anonymity is central to their function and mission. By avoiding the critical gaze of the authority they seek to subvert, they are able to challenge the power structure by operating in its very own structure. The collective identity of Anonymous is a challenge to the panoptic gaze of authority. As Carmela Ciuraru explains, "at its most basic level, a pseudonym is a prank."[86] Anonymous claims the subversive element of humor at the root of their mission that began with a series of hacks against the Church of Scientology in 2008, which its members claimed was done for "lulz," a term that appears in the group's *Encyclopedia Dramatica* as laughter directed at the victim of a prank.[87] Since then, their activities have evolved to challenge government agencies, copyright protection agencies, corporations, and child pornography sites. Anonymous expresses a sense of urgency in its mission statement: "We will either bring the whole system crashing down, starting anew community by community, or the current way will turn us all into serfs until humanity is more than decimated."[88]

Like occupation, anonymity when engaging the public sphere has a major role in the long history of democratic protest. Pen names and pseudonyms have been used to criticize power from below since the expansion of the public sphere in Renaissance Europe, which was aided by the modern forces of technology, market economies, and democratic ideology. Writers used pen names when their ideas were a threat to established order, so as to protect them from recrimination from authorities. Many editorial pieces in the seventeenth- and eighteenth-century English Atlantic world were written under the guise of pen names, and pen names were vital to the public discourse of the American Revolution. As Eran Shalev observes, pseudonyms were an important element in Revolutionary discourse because "a worthy cause would be better served by the reverberation of many voices, while maintaining the impression that the newspaper essays and pamphlets were spontaneous expressions of American public opinion. Thus, Benjamin Franklin recommended blanketing the colonies with anonymous and pseudonymous writings because they would 'render the discontents general . . . and not the fiction of a few demagogues.' Further, the use of pseudonyms indeed 'enabled men of honor to behave dishonourably.'"[89]

Similarly, the concealment of identity is elemental to the history of physical forms of public protest and sabotage. After the economic downturn caused by the South Sea Economic Collapse in 1720, groups of rural yeomen across the English countryside took to poaching as a means of subsistence. Parliament responded to these groups, called "The Blacks" because of the manner in which they painted their faces to conduct their raids at night, with a law called "The Black Act." Among the provisions of The Black Act was the condition that "anyone having his or their faces blacked, or being otherwised disguised . . . shall suffer death as in cases of a felony, without benefit of clergy."[90] Disguise was also an important element in the activities of the Sons of Liberty in the American colonies, from the razing of tax collectors houses to the destruction of the tea in Boston Harbor. The masks which Anonymous use in their communications are caricatures of Guy Fawkes, who plotted to destroy the House of Lords when the aristocracy and nobility were present to open Parliament in the Gunpowder Plot of 1605. The masks were popularized in the 2006 film *V for Vendetta*, which adapted the story for a modern audience. As James Fenimore Cooper accounted in his 1821 novel, *The Spy*, about the loyalties of citizens during the Revolutionary War in Westchester, New York, "great numbers, however, wore masks."[91] With the addition of virtual space as a second front in the struggle for human liberty, even greater numbers wear masks today.

Taken together, these two forms of resistance, Occupation and Anonymity, frame a means by which people can assert democratic ideology to resist the triangulation of physical and ideological space. In the process, they make the societies in which they manifest more inclusive, participatory, and just. Occupation is present, overt, and embodied, while Anonymity is obfuscated, hidden, abstract, and cryptic. Challenging the occupation of the public sphere by corporate power structures, and the anonymity that bureaucratic power structures provide the new global elite as a means to consolidate wealth and power by attacking them through the very means of their control, ontological stasis can be broken and teleological change can manifest in its place over time.

TOWARD THE TECHNOCRATIC MONOCULTURE

Within the context of modernity, these tensions between static hierarchical space and dynamic temporal change are both amplified and accelerated. Modernity itself is a theoretical construction, generally regarded as the third of three historical epochs, distinct from the ancient and medieval not as much in content, but in form and extent. Its principle characteristics are an exponential acceleration of technological capability, the expansion of the capitalist market economy, the dual forces of urbanization and nationalism, and the presence of democratic ideology. This is not to say that these elemental conditions were not present in the historical periods prior to the onset of modernity. Rather, the degree to which these conditional forms manifest or at least influence collective experience is generally recognized as more pronounced than in the ancient and medieval epochs.

The effect of this spatial triangulation is to minimize the possibility of teleological challenge to the dominant paradigm; it is predicated upon authority and control for the purpose of concentrating power in the hands of a relative few. This comes ultimately at the expense of the realization of democratic ideology, which is the teleological objective of historical phenomenology. All of these: technology, the dual forces of nationalism and urbanization, and the market economy, have been used to assert the hegemonic power of the capitalist nation-state at the expense of an ever-present, but ebbing and flowing, democratic ideology. Space, once triangulated from above, checks the teleology of phenomenological emancipation. Without space to operate in, democratic ideology, which ultimately leads to a return to commonwealth, has one metaphysical resource left: time. But teleology is interrupted by ontological spatial stasis, enforced by the comprehensive and expansive political, economic, and cultural force of the technocratic monoculture.

NOTES

1. Saint Augustine, *Confessions*, trans. J. G. Pilkington (Edinburgh: T.T. & Clark, 1876), Book XI, Chapter XIV, 301.
2. Ibid. Book XI, Chapter XXIII, 310.
3. Ibid. Book XI, Chapter XVIII, 304.
4. Hermann Minkowski, "Raum und Zeit." *Physikalische Zeitschrift*. Vol. 10, No. 104 (1909): 75–88.
5. Anthony Giddens, *A Contemporary Critique of Historical Materialism* (Berkeley and Los Angeles: University of California Press, 1983), 4.
6. Martin Heidegger, *Being and Time*, trans. Joan Stambaugh (Albany: State University of New York Press, 1996), 335.
7. George Friederich Hegel, *Hegel: The Essential Writings*, ed. Frederick G. Weis (New York: Harper & Row, 1974), 37–85.
8. Henri LeFebvre, *The Production of Space*, trans. Donald Nicholson-Smith (Malden & Oxford: Blackwell Publishing, 1991), 10–11.
9. Gerhard Dorn Van-Rossum, *The History of the Hour: Clocks and Modern Temporal Orders*, trans. Thomas Dunlap (Chicago and London: University of Chicago Press, 1996), 43.
10. Arno Borst, *The Ordering of Time: From the Ancient Computus to the Modern Computer* (Oxford: Polity Press, 1993), 104–105.
11. Laura Ost "NIST Ytterbium Atomic Clock Sets Record for Stablility," *NIST Tech Beat*, August 22, 2013, www.nist.gov/pml/div688/clock-082213.cfm
12. Richard Brautigan, *All Watched Over By Machines of Loving Grace* (San Francisco: The Communication Company, 1967).
13. Thomas Carlyle Smith, *Native Sources of Japanese Industrialization 1750–1920* (Berkeley: University of California Press, 1988), 204.
14. Brooke Hindle, *Emulation and Invention* (New York: NYU Press, 1981), 85.
15. Milan Kundera, *Slowness: A Novel*, trans. Linda Asher (New York: Harper Collins, 1997), 2.
16. Harlan K. Ullman and James P. Wade, Jr. *Rapid Dominance—A Force for All Seasons* (London: Royal United Services Institute For Defence Studies, 1998), 87.
17. G. Underwood and R. A. Swain, "Selectivity of Attention and the Perception of Duration." *Perception*, Vol. 2, No. 1 (1973): 101–105.

18. Rebecca Solnit, *Wanderlust: A History of Walking* (New York: Verso Books, 2002), 7.
19. Aldo Leopold, "Engineering and Conservation," in *The River of the Mother God, and Other Essays by Aldo Leopold*, ed. Susan L. Flader and J. Baird Callicott (Madison: University of Wisconsin Press, 1991), 254.
20. Jorge Luis Borges, "The Aleph," in *The Aleph and Other Short Stories*, trans. and ed. Norman Thomas di Giovanni (London: Jonathan Cape, 1970), 26–28.
21. Norman J. Thrower, *Maps and Civilization: Cartography in Culture and Society* (Chicago: University of Chicago Press, 2008), 26–33.
22. P.D.A. Harvey, *Mappa Mundi, The Hereford World Map* (Toronto: University of Toronto Press, 1996), 1–6.
23. Colonel Sir Charles Close, *The Early Years of the Ordnance Survey* (Devon: David and Charles Reprints, 1969), 16.
24. Eric G. Forbes, *Greenwich Observatory: One of Three Volumes by Different Authors Telling the Story of Britain's Oldest Scientific Institution* (London: Taylor and Francis, 1975), Vol. 1, 20.
25. Ibid. Vol. 1, 23.
26. Benedict Anderson, *Imagined Communities: Reflections on the Origin and Spread of Nationalism* (London: Verso, 1991), 6.
27. Ernest Gellner, *Thought and Change* (London: Weidenfield and Nicolson, 1964), 150.
28. Eddie Dixon, "Willie McCool Memorial," *City of Lubbock, Department of Parks and Recreation*, Accessed May 14, 2014, www.mylubbock.us/departmental-websites/departments/parks-recreation/top-navigation-menu-items/parks/attractions/willie-mccool-memorial
29. Eric Hobsbawm, *Nations and Nationalism since 1780* (Cambridge: Cambridge University Press, 1997), 14.
30. Josep Llobera, *The God of Modernity: The Development of Nationalism in Western Europe* (Oxford: Berg, 1994), 219.
31. Eric Hobsbawm, "Mass Producing Traditions: Europe, 1870–1914," in *The Invention of Tradition*, ed. Eric Hobsbawm and Terence Ranger (Cambridge: Cambridge University Press, 1996), 70–5.
32. Modris Eksteins, *Rites of Spring: The Great War and the Birth of the Modern Age* (New York: Mariner, 1989), 109–128.
33. Benedict Anderson, *Imagined Communities: Reflections on the Origin and Spread of Nationalism* (London: Verso, 1991), 163.
34. Eric Hobsbawm, *Nations and Nationalism Since 1780* (Cambridge: Cambridge University Press, 1997), 8.
35. Peter N. Stearns, *The Industrial Revolution in World History* (Philadelphia: Perseus, 2013), 32.
36. E. P. Thompson, "Time, Work-Discipline and Industrial Capitalism." *Past and Present*, No. 28 (December 1967): 69.
37. Jacques Le Goff, *Time, Work and Culture in the Middle Ages* (Chicago: University of Chicago Press, 1980), 31–37.
38. D. Brendan Nagle and Stanley M. Burstein, *The Ancient World: Readings in Social and Cultural History* (Englewood Cliffs: Prentice Hall, 1995), 30–31.
39. Michael Biggs, "Putting the State on the Map: Cartography, Territory and European State Formation." *Comparative Studies in Society and History*, Vol. 41, No. 2 (April 1999): 385.
40. Ambrose Bierce, *The Unabridged Devils Dictionary*, ed. David E. Schultz and S. J. Joshi (Athens: University of Georgia Press, 2000), 28.
41. Karl Marx, *Capital: A Critical Analysis of Capitalist Production* (London:, Swann & Sonnenschein & Co., 1902), 748.

42. Joseph P. Schweiterman and Dana M. Caspall, *The Politics of Place: A History of Zoning in Chicago* (Chicago: Lake Claremont Press, 2006), 17–34.
43. Carl Chinn, *Poverty amidst Prosperity: The Urban Poor in England, 1834–1914* (Manchester: Manchester University Press, 1995), 76–79
44. Kevin Petersen, "Room To Grow: Detroit Takes First Steps to Legalize Urban Agriculture," *Michigan Journal of Environmental and Administrative Law*, February 8, 2013. http://students.law.umich.edu/mjeal/2013/02/room-to-grow-detroit-takes-the-first-steps-to-legalize-urban-agriculture/
45. City of Gary Indiana, "Dollar Home Project," Accessed May 14, 2014. www.teamgaryindiana.com/?p=1232
46. Hesiod, "Works and Days," in *Hesiod and Theogonis*, trans. Dorothea Wender (New York: Penguin Books, 1973), 71.
47. Aristotle, *Politics*, trans. Benjamin Jowett (Oxford: Clarendon Press, 1905), Book I, 46.
48. T. G. Tucker, *Life in Ancient Athens* (London: Macmillan, 1907), 81.
49. T.B.L. Webster, *Everyday Life in Classical Athens* (New York: G.P. Putnam, 1969), 61.
50. Ramsay MacMullen, "Market-Days in the Roman Empire." *Phoenix*, Vol. 24, No. 4 (1970): 334.
51. Macrobius, *Saturnalia* (Boston: Loeb Classical Library, 2011), 162–77.
52. Jacques Le Goff, *Time, Work and Culture in the Middle Ages* (Chicago, University of Chicago, 1980), 29.
53. Samantha Letters, *Gazetteer of Markets and Fairs in England and Wales to 1516* (London: Centre for Metropolitan Research, 2004). Last Modified Dec. 16, 2013, www.history.ac.uk/cmh/gaz/gazweb2.html
54. Exodus 20:8–11, *The Bible: Authorized King James Version*, ed. Robert Carroll and Stephen Prickett (Oxford: Oxford University Press, 2008), 90.
55. Alvin W. Johnson and Frank H. Yost, *Seperation of Church and State in the United States* (Minneapolis: University of Minnesota, 1948), 220.
56. Ibid. 221.
57. Ibid. 224.
58. *Nasdaq Regular Trading Session Schedule*, accessed May 27, 2014, www.nasdaq.com/about/trading-schedule.aspx
59. Douglas McIntyre, "24 Hour Stock Trading," *24/7 Wall Street*, February 15, 2013, http://247wallst.com/investing/2013/02/15/24-hour-stock-trading/
60. Brandon Keim, "Nanosecond Trading Could Make Markets Go Haywire," *Wired Magazine*, February 16, 2012, www.wired.com/wiredscience/2012/02/high-speed-trading/
61. Stephen Duck, "The Thresher's Labour," in *Poems for Several Occasions* (London: John Osborn, 1738), 12.
62. Henri LeFebvre, *The Production of Space*, trans. Donald Nicholson-Smith (Malden & Oxford: Blackwell, 1991), 10.
63. Karl Polanyi, "Ports of Trade in Early Society," in *Primitive, Archaic and Modern Economies*, ed. George Dalton (New York, Doubleday, 1968), 2389.
64. Douglas M. McDowell, *The Law in Classical Athens* (London: Thames and Hudson, 1978), 157.
65. Henri LeFebvre, *The Production of Space*, trans. Donald Nicholson-Smith (New York: Blackwell, 1991), 8.
66. R. H. Hilton, "Small Town Society in England before the Black Death," in *The Medieval Town 1200–1540*, ed. Richard Holt and Gervase Rosser (New York: Longman, 1990), 75.
67. David Nicholas, *The Growth of the Medieval City, from Late Antiquity to the Early Fourteenth Century* (Harlow: Addison Wesley Longman, 1997), 114.
68. Robert M. Fogelson, *Downtown: Its Rise and Fall 1880–1950* (New Haven: Yale University Press, 2001), 12–3.

69. Henri LeFebvre, *The Production of Space*, trans. Donald Nicholson-Smith (New York, Blackwell, 1991), 36.
70. Christian Parenti, *The Soft Cage: Surveillance in America, From Slavery to the War on Terror* (New York, Basic Books, 2004), 5.
71. Douglas Reid, "The Decline of Saint Monday 1766–1876." *Past and Present*, Vol. 71, No. 1 (May 1976): 100–101.
72. Rowland Hill Harvey, *Robert Owen: Social Idealist* (Los Angeles: University of California Press, 1949), 193–194.
73. Jonathan Grossman, "Fair Labor Standards Act of 1938: Maximum Struggle for a Minimum Wage." *Monthly Labor Review*, Vol. 101, No. 6 (June 1978): 22.
74. Erik Rauch, "Productivity and the Workweek," accessed May 28, 2014, http://groups.csail.mit.edu/mac/users/rauch/worktime/
75. Anna Coote, Andrew Sims, and Jane Franklin, *21 Hours*, New Economics Foundation, February 13, 2010, www.neweconomics.org/publications/entry/21-hours
76. Julia C. Bishop and Mavis Curtis, *Play Today in the Primary School Playground* (Buckingham: Open University Press, 2001), 183.
77. Michael Kazin and Steven J. Ross, "America's Labor Day: The Dilemma of a Workers' Celebration." *The Journal of American History*, Vol. 78, No. 4 (Mar. 1992): 1321.
78. Emma Brown, "D.C. Parents Push For More Recess," *The Washington Post*, August 30, 2013, http://articles.washingtonpost.com/2013–08–30/local/41600 378_1_montgomery-county-schools-recess-minimum
79. Michael Pollan, *Food Rules: An Eater's Manual* (New York: Penguin Books, 2009), 55.
80. "Holidays, Time off, Sick Leave, Maternity and Paternity Leave," *Working, Jobs and Pensions*, accessed May 27, 2014, https://www.gov.uk/browse/working
81. Sylvia Moore, "Occupy Wall Street Movement on Health Care," October 15, 2011, http://californiaonecare.org/occupy-wall-street-movement-on-health-care/
82. "Occupy Olympus at NY International Fringe Festival 2013," accessed June 24, 2014, http://magistheatre.org/current_production.html
83. Lizzie Davies, "Occupy Movement: City-By-City Police Crackdowns So Far," *The Guardian*, November 15, 2011, www.theguardian.com/world/blog/2011/nov/15/occupy-movement-police-crackdowns?CMP=twt_gu
84. R. Develin. "'Provocatio' and Plebiscites. Early Roman Legislation and the Historical Tradition." *Mnemosyne*, Fourth Series, Vol. 31, Fasc. 1 (1978): 45–60.
85. Le Corbusier, *The Modulor*, trans. Peter de Francia and Anna Bostock (Basel: Birkhauser Architecture, 2000), 30.
86. Carmela Ciuraru, *Nom De Plume, A (Secret) History of Pseudonyms* (New York: Harper Collins, 2011), xiii
87. "Lulz," *Encyclopedia Dramatica*, accessed October 15, 2013, https://encyclopediadramatica.es/Lulz
88. Anonymous, *United Mission Statement*, accessed October 15, 2013, http://anoncentral.tumblr.com/post/11964602984/the-text-of-the-well-received-flier-being-passed-around
89. Eran Shalev, "Ancient Masks, American Fathers: Classical Pseudonyms during the American Revolution and Early Republic." *Journal of the Early Republic*, Vol. 23, No. 2 (Summer 2003): 158.
90. E. P. Thompson, *Whigs and Hunters: The Origin of the Black Act* (London: Penguin Books, 1975), 271.
91. James Fenimore Cooper, *The Spy: A Tale of Neutral Ground* (New York: W.A. Townsend and Company, 1861), 12.

3 Technocratic Monoculture

INTRODUCTION

The overarching effect of temporal and spatial control imposed from above in the age of modernity is the spread of technocratic monoculture. Here, the forces of technology, nationalism and urbanization, and market economics work together to challenge and negate the presence of democratic ideology and the possibility of teleological progress in time through the hierarchical enforcement of ontological stasis in space. The degree to which these forces have accelerated over the course of modernity have enabled the technocratic monoculture to become more pervasive and complete. As The Invisible Committee, the anonymous authors of the 2007 critique of global capitalism *The Coming Insurrection*, suggest: "As an attempted solution, the pressure to ensure that *nothing happens*, together with police surveillance of the territory, will only intensify."[1] This hierarchically imposed and enforced stasis, *to ensure that nothing happens*, is both temporal and spatial in design. The result is a feedback loop of superficial change without any structural shift to the underlying order in which those changes take place.

Stasis is enforced spatially through the imposition of a comprehensive system of surveillance constantly expanding its reach to observe and account for everything that happens in the space it monitors. Following the imposition of surveillance, anything that occurs within that space can then be commodified. That is to say it can be stripped of its intrinsic value in exchange for economic value. Once commodified, the space and the things that occupy it (resources, people, and ideas) can be controlled, manipulated, and directed toward the goal of maintaining and reinforcing the very system in which they exist. As Fredy Perlman observed in *Against-His-Story, Against Leviathan!*, "The Leviathan is a cannibal." It is expansive, singular and comprehensive. "Its enemy is everything outside of itself."[2] Where Perlman used the conceptual metaphor of the Leviathan to explain the origins and characteristics of civilization, its modern manifestation within the context of modernity is technocratic monoculture.

The term "technocratic monoculture" first appears in a 1991 essay by James Clifford entitled *Traveling Cultures*, in which he suggests that travel and movement promotes "discrepant cosmopolitanisms," which balance the dual forces of "excessive localism of particularist cultural relativism, as well as the overly globalist vision of a capitalist or technocratic monoculture."[3] Alan R. Drengson expanded on the idea, albeit indirectly, in his 1995 essay *Shifting Paradigms: From Technocrat to Planetary Persons*, in which he argued that "the technocratic state emphasizes wealth, power and the capacity to influence others. . . . The forms of organization that arise are all corporate entities, whether business, university, government, or military. These all converge in the technocratic state. Diversity is discouraged, for such a state tends to become a monolithic monoculture."[4] Since Clifford's initial observation and Drengson's exegesis, however, the concept has remained unelaborated and static. While Clifford's term is often cited in relation to connected topics, the term "technocratic monoculture" has not been critically developed as a condition relative to the subject of modernity. As such, it is a term open to exploration and analysis.

This chapter is a consideration of the concept of technocratic monoculture. First by considering the roots of the concept, then by examining its component parts, technocracy and monoculture, it seeks to establish a working understanding of the term. From there, it explores how technocratic monoculture has been imposed spatially over time and has resulted in a discourse of security that undermines freedom and enforces assimilation, and in the process enforces ontological stasis to prevent the manifestation of teleological change. To the initial point of consideration, it would seem, ultimately, that technocratic monoculture as an overarching condition of modernity is essentially a whole of two parts: the prison-industrial complex and market economics.

SURVEILLANCE AND COMMODITY

The prison-industrial complex enforces an autonomous hierarchical order of both time and space. Incarceration is the ultimate form of temporal and spatial control, whereby the entire sensory experience of those subject to imprisonment is designed, ordered, and regulated by a hierarchical authority. The phrase "prison-industrial complex" was first coined by the social justice activist Angela Davis in 1997 as a variation on the phrase "military-industrial complex." It is a business that produces exponential profits, and whose main product is systematic oppression. Davis articulated the comprehensive nature of the project:

> But private prison companies are only the most visible component of the increasing corporatization of punishment. Government contracts to build prisons have bolstered the construction industry. The architectural community has identified prison design as a major new niche. Technology developed for the military by companies like Westinghouse is being marketed for use in law enforcement and punishment.[5]

The origins of the prison industrial complex can be found at the very foundations of transatlantic modernity, at the juncture between central authority and peripheral control. It was during this time that reformers looked to the prison system in hopes of reasserting control over a proletarian population that had gotten out from under the control of hierarchical authority in an age where the new governing systems of modernity, driven by capital accumulation, were replacing the old medieval systems, driven by religiosity and dogma. The eighteenth-century English prison reformer John Howard observed that the opportunity existed to promote "the more rational plan for softening the mind in order to its amendment" by shifting personal control of prisoners by gaolers, which could be arbitrary and inconsistent, with a more systematic design for the prison itself.[6] The new architecture that resulted from these prison reforms proved so effective that their design and intent became a model for society itself. If prisoners could be controlled and "reformed" to defer to the hierarchical order of the new model, so too could citizens at large. Howard's model proposed that regulations governing security, health, diet, clothing, lodging, firing (heat), religious instruction and morals, employment, rewards, punishment, treatment of sick, and proceedings on death all be established as the parameters with which to govern prison life. The result of such comprehensive institutional regulation and control would be to "correct the faults of prisoners, and make them for the future useful to society."[7] The application of the modern prison to the landscape of global capitalism became a vital tool in the command and control of space in the early modern world, and its continuous refinement is something that continues through to the present.

The design of physical space for the purpose of hierarchical surveillance and control as a central function of prison architecture was effectively reified by Jeremy Bentham, the eighteenth-century political economist who laid the architectural and ideological plans for the Panopticon. Where Howard advocated the dismantling of inmate community through isolation, Bentham contributed to this concept by subjecting the prisoner to the constant observation of the gaoler through the functional design of the prison. The preface to Bentham's plan for the Panopticon, published in 1787, lauds its virtues accordingly: "Morals reformed—health preserved, industry invigorated, instruction diffused-public burthens lightened—Economy seated, as it were, upon a rock-the Gordian knot of the Poor Laws are not cut but untied-all by a simple idea in Architecture!"[8] In the Panopticon, spatial control lays a foundation for ideological control. The ability of the inspector, who sits at the center of the building, to effect control of the inmates who surround him on the periphery, rests on his ability of "seeing without being seen." The architectural design of Bentham's Panopticon makes it so that "the persons being inspected should always feel themselves as if under inspection, at least as standing a great chance of being so."[9]

This "apparent omnipresence of the inspector" serves a number of functions. There is efficiency to the design. One overseer can monitor many subjects from his central location. The design of seeing without being seen is intended to bring the subjects to internalize the gaze of the overseer. By never knowing if he or she is being watched, but assuming the possibility of it, the subject will behave according to the expectations of the overseer watching him or her. Once incarcerated, the prisoners could be engaged to participate in "pecuniary economy." In this step, the prison becomes a factory, to promote a capitalist mode of production. Here, Bentham proposed that a "contractor... With his promising hands and his drones... would set up a manufacture."[10] The Panopticon as both prison and factory was also a model that Bentham applied to schools. As an educational model, Bentham distinguished between "two very distinguishable degrees of extension:—It may be confined to the hours of study; or it may be made to fill the whole circle of time, including the hours of repose, and refreshment, and recreation." The origins of Bentham's inspection house lay in the prison reform movement of the eighteenth-century Atlantic world, but its institutional efficacy in command and control through spatial design resonate outside of those temporal and spatial parameters.

The lack of privacy prisoners have in the Panopticon renders them vulnerable to the disciplinary scrutiny of the prison overseer, who has complete privacy. This dynamic is elemental to the establishment and proliferation of control inside its walls. As Michel Foucault wrote:

> A whole problematic develops: that of an architecture that is no longer built simply to be seen, (as with the ostentation of palaces), or to observe the external space (cf. the geometry of fortresses), but to permit an internal, articulated and detailed control—to render visible those who are inside it; in more general terms, an architecture that would operate to transform individuals: to act on those it shelters, to provide a hold on their conduct, to carry the effects of power right to them, to make it possible to know them, to alter them.[11]

Here, secrecy and exposure are bound up in the power dynamic of the panoptic model of spatial and ideological control. This is an essential component of the technocratic monoculture. The state, and the corporations who fund it, both insist on the necessity of access to panoptic surveillance. It is critical to their functionality, while the reversal of the optic is criminalized: *scene* and *obscene*. Privacy, articulated both legally and morally as modern democratic right, becomes privatization, an economic privilege within the context of technocratic monoculture.[12] Privacy is a right of the state and the corporations that fund it as an extension of their interests, while constant and ever expansive surveillance systems monitor the physical and ideological activities of those trapped within its political, economic, and social structures.

The Panopticon is, above all, an economic institution. Its principle aim, as Bentham designed it, is to "obtain power of mind over mind" for "the joint purposes of punishment, reformation and pecuniary economy."[13] Management of the inspection house, whatever form it took, was to be done by contract. "I would farm out the profits, the no-profits, or if you please the losses, to him who, being in other respects unexceptionable, offered the best terms."[14] This new model of hierarchical social organization was based, above all, on capitalist enterprise. The main motive behind the inspection house is the extraction of wealth for a few based on the labors of the many. To ensure that those labors are maximized, Bentham proposed a model of top-down surveillance along with a system of rewards and punishments for those subject to its terms and conditions.

The second element of technocratic monoculture, then, is market economics. This system allows for the appearance of freedom which obfuscates the nature of hierarchical control in which it operates. Choices within this system are superficial, and firmly controlled by the architects of the temporal and spatial system in which they manifest. Andrew Bard Schmookler explains the illusion of choice in a capitalist system: "over time, the [market] system, because of its biases and distortions, carries us to a destination chosen by that system and not by us."[15] Any choice allowed within the panoptic model is artificial, a Morton's Fork which ultimately confirms the power dynamic of hierarchical control. Here, the market reinforces the authoritative nature of the prison-industrial complex. Together, both become the enabling mechanisms for the comprehensive nature of technocratic monoculture. Political and social choices are appropriated by the market, which makes every choice an economic choice.

Herein, choice is commodified and stripped of political and social meaning. The spatial expansion of markets and the construction of time as an economic resource together create a comprehensive system, which is increasingly difficult to subvert, undermine, or escape. The more ubiquitous the market becomes, the more it imposes a uniform condition on its subjects. As David Levine put it: "it is the genius of modern capitalism that producers must also be consumers, and in precisely this way the fetters of production have been replaced by those of consumption. Most members of modern society, like Gulliver awakening in Lilliput, have been tied to the consumer economy by a thousand tiny chains."[16] The system that subjects people to a model of hierarchically ordered economic productivity is reinforced by the system that restricts them to hierarchically ordered economic consumption. This system proliferates of its own accord, becoming increasingly woven in to the temporal and spatial fabric of modernity. The incarceration of territories and populations within the bounds of a singular market is the essence of technocratic modernity. Its expansion is something that the political scientist Benjamin Barber has described as "the integrative modernization and aggressive economic and cultural globalization" which he calls "McWorld," due to the essentially western origins of its institutions.[17]

Central to the expansion of markets is the idea of commodification. "Commodities" is the first subject in Marx's critical analysis of capitalism:

Das Kapital. The very first sentence of the work reads: "The wealth of those societies in which the capitalist mode of production prevails, presents itself as 'an immense accumulation of commodities.'"[18] Commodification is essentially the assignment of economic value to any particular thing. The expansion of markets is inherently bound up with the expansion of commodification. Over the course of history, markets have expanded both territorially and ideologically to become increasingly comprehensive. The Agricultural Revolution which marked the shift from the Paleolithic to the Neolithic Age allowed for the commodification of land as a principle resource for food production. This was followed by the Industrial Revolution and the Age of Machines, which enabled the increased commodification of raw materials necessary for production. The Digital Revolution, which spawned the Information Age, has since expanded the market to ideas and information, which are now commodified, bought, and sold as so many acres of land or timbers of oak. Throughout these developments, the commodification of space and time has been integral to the systematic expansion of the technocratic monoculture. Within the economization of these elemental forms, the commodification of everything that manifests in them, both physically and metaphysically, becomes possible.

The nature of commodification within the scope and context of modernity is as comprehensive as it is elemental. The first commodified element was the earth with the enclosures movement, which laid the foundation of the shift from the feudal to the industrial economy. The next commodified element was fire. With enclosures came the refusal of rights of estover, which provided the custom of firewood for manorial laborers. Then came the commodification of the subterranean hydrocarbons, coal, oil, natural gas, which fuelled the successive waves of industrialization with resources that provided fire from below the earth's surface. This was evocative of an arrangement with hell itself, as with Blake's reference to the "Satanic Mills" during the onset of the Industrial Revolution in the nineteenth-century English midlands. From hell below to heaven above, air was the next commodified element. With the burning hydrocarbons, the expansive factory system continue to pollute the atmosphere, and laws to promote clean air have been met with stiff resistance by industries more concerned with profits than planetary biospheric integrity. The ultimate consummation of the political and the economic came with the Orwellian-named "Clear Skies Initiative of 2003," which actually increased the amounts of pollutants that American manufacturers could release into the atmosphere.[19] Finally, the elemental cycle of commodification is nearly complete, with the commodification of water being the last great hurdle for the agents of capitalism to overcome.

Increasingly, more corporations are building and controlling the infrastructure of water distribution, replacing public control with private interest. As of 2010 to 2011, 12% of the world population's access to water was controlled by corporate interests.[20] A 2009 report titled *A Concise Review of Challenges and Opportunities in the World Water Market* put forth by The TechKNOWLEDGEy Strategic Group observes that "there has been a surge of financial interest—from all types of investors—in the water business," and assures potential investors that "water prices *will* rise, and over time water *will* come to be viewed more and more as a true economic commodity."[21] At the same time, Nestle Chairman Peter Brabeck-Letmathe has argued that "access to water is not a public right," and that water should be privatized and priced at market value.[22]

David Harvey explains that "commodification presumes the existence of property rights over processes, things and social relations, that a price can be put on them, and that they can be traded subject to legal contract. The market is presumed to work as an appropriate guide—an ethic—for all human action."[23] The bounds of the market place are blurred and erased in the technocratic monoculture. The market becomes the preeminent governing tool, which is of course predicated on a financial ability to participate in such a governance model. Supreme Court decisions asserting that corporations are people and money is talk that exacerbates the inequality inherent in this model, as do World Trade Organization policies that promote the economic interests of corporations and nation-states without regard for the global citizens who are directly impacted by the economic and environmental fallout that results from such unilateral assertions.[24]

The obliteration of choice outside of the sanctioned parameters of the technocratic monoculture, combined with the commodification of everything within its bounds, ultimately renders those subject to its conditions powerless. As Janet Semple explains, "the inmate of the panopticon is deprived of choice, indeed in the pauper panopticons brought up in ignorance of the outside world, deliberately deprived of the knowledge that would enable him to make a rational choice in accord with his own interest."[25] Concurrently, the physical and ideological space to resist the encroachment of the market economy are policed, monitored, and controlled with increased technological precision. Choice, reduced to superficial options that contribute to the overall governing autonomy of the market, is stripped of any political or social implications. Space and ideology are subject to observation; time and activity are subject to regulation. Again, with Foucault, "the time table is an old inheritance. The strict model was no doubt suggested by the monastic communities. It soon spread. Its three great methods—establishing rhythms, impose particular occupations, regulate cycles of repetition—were soon to be found in schools, workshops and hospitals."[26] The technocratic monoculture becomes a metaphysical manifestation within the bound and regulated confines of panoptic control.

From the imposition of the panoptic model of hierarchical control to the systematic expansion of the market economy within it, technocratic monoculture becomes an expansive and comprehensive system. It is a product of thoroughly modern design, which makes use of all the key aspects of modernity to expand and become increasingly more pervasive. Technology becomes the means by which the surveillance apparatus of the corporatized state expands to observe both the physical and ideological activities of its subjects. The dual forces of nationalism and urbanization provide fixed markets for the institutionalization and expansion of market economies, rendering its subjects unable to contest the localized terms and conditions imposed by transnational, global capital. Market economies expand through the commodification of everything within the panoptic view, both physical and ideological. Within this framework, democratic ideology is marginalized and problematized. Citizens are transformed into consumers; active participants engaged in political and cultural processes are reduced to passive recipients of economic processes, which are imposed from above. The design is, ubiquitous and constantly reinforced, replicated in nearly every public and private institution within the scope of the overseer's view.

To understand the ways technocratic monoculture imposes an ontological stasis that prevents any kind of teleological progress from occurring, we should examine its component parts and see how they come together in systematic ways. First, with a consideration of the modern phenomenon of technocracy, then by examining the concept of monoculture, and finally, by considering the ways that together, technocratic monoculture has developed over the course of history to, in the words of the anonymous Invisible Committee, "*ensure that nothing happens.*"[27]

TECHNOCRACY

Technocratic Monoculture is a fusion of two concepts: Technocracy and Monoculture. To understand the concept, we should first isolate its elements and consider each on their own terms. Technocracy is a concept rooted in late nineteenth-century intellectual thought. Early uses of the term are scattered throughout art, geographical, historical, and economic journals, with no consistent application or theory behind it. *The Art-Journal*, a London-based publication, was one of the earliest to commit the word to print in 1855, warning that "great revolutions are effected [sic] in our time. The free and easy republicanism of Art has departed, and we may soon live under an iron technocracy destructive alike of both eyesight and nerves."[28] The concept references both artistic aesthetic and political theory. There is no elaboration on the concept; the author of the article is not even credited, yet the term is evocative. We can consider its evolution over the course of modernity from a term of critical appraisal to one of political liberation, and its eventual connotation as a term of comprehensive oppression.

The term technocracy assumed political and economic implications in 1919, with the publication of William Henry Smyth's article *Technocracy—National Industrial Management*. In the wake of The Great War, Smyth observed that the United States had become "a real Industrial Nation." This was the result of the organization and coordination of "the Scientific

Knowledge, the Technical Talent, the Practical Skill and the Man Power of the entire Community: focusing them in the National Government, and applying this Unified National Force to the accomplishment of a Unified National Purpose." Smyth asserted that "for this unique experiment in rationalized Industrial Democracy I have coined the term 'Technocracy.'"[29] In the subsequent essay, *Ways and Means To Gain Industrial Democracy*, Smyth explained that "when we speak of Industrial Democracy, what we really mean is: Nation-wide Industry managed by Technologists—a Nation of free and socially equal workers, scientifically organized for mutual benefit and unified purpose—a Technocracy."[30]

While Smyth's vision of technocratic future were inherently nationalist and utopian, the American economist Thorstein Veblen exposed and articulated the economic interests of technocratic ideology in his 1921 work, *Engineers and the Price System*. Veblen argued that the origins of entrepreneurialism "are both technological and commercial," but that the growth of both interests as technical expertise and "corporation finance" capital expanded "has brought on a further division, dividing the ownership of the industrial equipment and the resources from their management. But also at the same time the industrial system, on its technological side, has been progressively growing greater and going farther in scope diversity, specialization and complexity . . ."[31] Put plainly, the interests of both technology and capital were eclipsing the interests of the people that technocracy, or industrial democracy, was intended to serve. According to Veblen, technocracy had been appropriated by commerce, and industrial democracy was an impossibility because "in the American frame of mind absentee ownership is the controlling center of all the economic realities."[32]

Veblen's critique notwithstanding, technocratic philosophy reached something of a popular peak in the early 1930s, at the height of the Great Depression. In 1933, with unemployment at 25% in the United States, technocracy's chief public advocate, Howard Scott, a chemical engineer whose socially progressive views brought the patronage of the Industrial Workers of the World earlier in his career, articulated a vision of society which promised that "for the first time circumstances will make it possible for us to live without the entrepreneur." Among the things Scott called for in the technocratic platform were the abolition of money (to be replaced by the measure of energy any individual consumed over a given period of time), the four hour workday (four days a shift, 165 days a year), and an "Era of Peace and Plenty."[33] It was a message that resonated for a time in the public sphere, but was just as quickly dismissed by its institutional targets. In 1933, Benjamin DeCassares suggested in *Life* magazine that "the Idea Racketeers have invented Technocracy," and that "they are the new Wisenheimers of the race."[34] John Langdon-Davies observed in *Forum and Century* that "it is unfortunate that the collapse of technocracy under the strain of stupid publicity and overstatement has buried beneath its ruins several fascinating questions of first rate importance." Ultimately, Davies concluded that the questions at the source of technocratic ideology were best answered by the market economy that it criticized.[35]

The technocracy of Howard Scott and his contemporary Harold Loeb advocated a society organized by engineers to serve the public interests, as the model of a society organized by the interests of capital only served private interests. The group became splintered on principles. Loeb went on to found the Continental Committee on Technocracy in 1933, which was driven by an element of social justice.[36] Scott's group, Technocracy, Inc., evolved along more fascistic lines. In his self-published essay *Technocracy: Science vs. Chaos*, the cover of which features a stylized version of Zeus discharging lightning bolts from his fists (See Figure 3.1), Scott's plan for the new Energy Economy was totalitarian:

> The only way for an individual under Technocracy not to participate in this income would be:
> 1. To leave the Continent permanently.
> 2. To commit suicide.
> 3. To induce the state to execute him.[37]

Such discourse alienated more people than it engaged. Ultimately, the Great Depression gave way to the war economy and the New Deal, and technocratic ideas, always based on somewhat utopian models and theories, were lost to the more pragmatic and institutional status quo. By 1939, considerations of technocratic theory as a way to challenge the compound problems of modernity were essentially relegated to the more eccentric margins of intellectual discourse.

While discussions about technocracy declined with the New Deal and the beginning of World War II, the figure of the technocrat emerges as a predominant term in political discourse beginning in the late 1950s. As the postwar economies took shape in both America and Europe, the term seems to have evolved out of the need for the state to manage the infrastructural resources of the commercial sector in order to render market economies effective in the fight against communism. Here, expertise was marshaled to reinforce market economies, not to challenge or subvert them. In observing the economic blockade that prevented businesses in allied controlled parts of Europe and Asia from trading with the Soviets, C. L. Sulzberger asked in a 1952 article: "Can the technocrat democracies of the West ignore indefinitely such Iron Curtain raw materials as manganese and monazite sands?"[38] Technocrats would need to figure out a way to promote market capitalism, rather than subverting it. By this point, the term had been effectively appropriated by business interests and stripped of its original meaning and intent.

The use of the term "technocrat" to mean an expert employed by the state to promote the interests of capitalist market economics exploded during the 1960s, as the Cold War with the Soviet Union intensified while the terms

Figure 3.1 Science vs. Chaos by Howard Scott, 1933.
https://archive.org/details/ScienceVsChaos. The Internet Archive.

capitalism and democracy became interchangeable in the fight against communism. A *New York Times* article reporting on the economic redevelopment of Newcastle-Upon-Tyne in England observed that, "the technocrats, backed up by the politicians . . . drew up an ambitious urban redevelopment plan in 1963. . . ." In the plan, "40% of Newcastle's homes either will be pulled down or replaced by new housing" and "will give way to a modern two-tiered shopping and business center with the traffic area below and the pedestrian area above."[39] In Africa, Lloyd Garrison observed that "a revolution of the technocrats" in Nigeria, Dahomey, the Central African Republic, and Ghana, had resulted in "dependence on Peking and Moscow [being] reversed overnight," and had given "private enterprise . . . a chance to breathe."[40] By the 1970s and 80s, the term was ubiquitous, employed by journalists to identify anyone within the political bureaucracy whose opinion informed policy.

The legal apparatus of the business class has expanded exponentially in the 1980s under the Reagan administration to the point of subsuming all other interests in the field of politics. Capital had hijacked the political framework of representative government to promote its interests over those of the collective body it governs. Expertise was politicized. The result is what the political scientist Frank Fischer calls "The New Class," which is essentially the rise of policy experts funded by business interests to provide the superficial appearance of technocratic expertise. Their neoconservative agenda, Fischer asserts, "threatens the future of representative government" by seeking to systematically exclude the body politic from political decision making and promoting the consolidation of wealth and power through an ideological construct of technical authority and false impartiality.[41]

Here, the veneer of technocracy, based on the idea that experts are in the best position to determine the solutions of complex problems, operates to disenfranchise the citizenry to the benefit of the business interests who use political means to achieve advantageous economic ends. From this argument, the rise of the modern technocrat, policy experts on everything from education to criminology to the Middle East and Meteorology, are essentially bought and paid for by business interests to promote "market-based solutions" to social and political problems caused, in large part, by the very maldistribution of wealth that the modern technocracy promotes.

Whereas the idea of technocracy as put forth by William Henry Smyth in 1919 was essentially political and economic, *The Art-Journal*'s original application of the term, which predates Smyth's by sixty-four years, is cultural. The observation that technocracy is "destructive alike of both eyesight and nerves" is a reference to the sensory effects that technology has on the aesthetics of art and life. In Victorian England, the notion that modernity was inherently offensive to the senses was a matter of medical opinion. Male doctors often treated women who could afford medical attention for "the modern conception of hysteria," which was believed to be associated with "exposure to the stresses of life."[42] Those who could not afford such attention were left to suffer the compounded sensory assault of industrial modernity without recourse.

Factories were new human environments at the onset of the Industrial Revolution, and their effect on the senses was comprehensive. The scale of the new factories led Christian P. W. Beuth, a German politician touring England in 1823, to comment that "the modern miracles, my friend, are to me the machines here and the buildings that house them, called factories." Upon observing the boiler house chimneys he observed that "it is hard to imagine how they remain upright."[43] The noise inside of factories was overwhelming. Lucy Larcom, who worked at the Boott Mill in Lowell, Massachusetts wrote in her 1889 autobiography that "the noise of the machinery was particularly distasteful to me . . . I discovered, too, that I could so accustom myself to the noise that it became like a silence to me. And I defied the machinery to make me its slave."[44] Like the noise, the temperatures in which industrial labor was conducted could be oppressive. The twelve-year-old coal bearer Isabella Read testified to the Lord Ashley's Mines Commission in 1842 that "when the weather is warm there is difficulty in breathing."[45] The air was toxic. In St. Helens, "the dense acidic smoke from the alkali works" created an "acrid, evil smelling environment," darkening the city with its "black clouds of smoke that rolled out of the factory chimneys."[46] All of this brought the surgeon John Malyn to testify before Parliament in 1832 that "the irritation thus produced might render the nervous system more sensitive in general."[47]

Industrial cities, too, were new human environments, and in them modernity's assault on the senses was compounded. Air, light, and noise pollution created new conditions with which humans had to learn to adapt to. Cities were dirty places, filled with all forms of biological and industrial waste. Freiderich Engels described the streets of Manchester in 1844 as "a state of filth! Everywhere heaps of debris, refuse, and offal; standing pools for gutters, and a stench which alone would make it impossible for a human being in any degree civilised to live in such a district." Animal waste contributed to the stench and the filth: "Here, as in most of the working-men's quarters of Manchester, the pork-raisers rent the courts and build pig-pens in them. In almost every court one or even several such pens may be found, into which the inhabitants of the court throw all refuse and offal, whence the swine grow fat; and the atmosphere, confined on all four sides, is utterly corrupted by putrefying animal and vegetable substances. . . ." Along with these came industrial waste. "Above the bridge are tanneries, bone mills, and gasworks, from which all drains and refuse find their way into the Irk, which receives further the contents of all the neighbouring sewers and privies." All of this led Engels to conclude that "everything which here arouses horror and indignation is of recent origin, belongs to the industrial epoch."[48]

As civilian life became more technocratically modern with the onset of industrialization and urbanization, so too did military life. The Great War which began with the pageantry of marching formations and cavalry charges in 1914, ended in 1918 with gas attacks, tanks, trench combat, artillery bombardment, aerial warfare, and submarine warfare. Doctor Charles Myers, a captain in the Royal Army Medical corps, observed that the overwhelming nature of technological warfare had a deleterious effect on soldiers' senses, often traumatizing them to the point of incapacity. Myers catalogued the effects of war on soldiers' sense of sight, hearing, smell, taste, touch, along with its effects on memory, excitability, sleep patterns, concluding that "the close relation of these cases to those of 'hysteria' appears fairly certain." He titled his study, published in the British medical journal *The Lancet*, in 1915: "A Contribution to the Study of Shell Shock."[49] While the physical effects of war had been understood throughout the history of warfare, the psychological effects of war were just beginning to be understood. As warfare changed and evolved within the context of modernity, becoming more technological and more comprehensive in scope, so too did medical understandings of its effects.

Technocracy continued to evolve over the course of the twentieth century. As the rise of the factory system in the Atlantic world receded, new types of work created new working environments. Factory work gave way to office work. Technocracy, however, remained a constant throughout the transition from industrial to post-industrial economies. The phenomenon of "technostress," a term that emerged as computers were becoming increasingly integrated into all facets of modern life in the latter part of the twentieth century, suggests the continued presence of the cultural conception of technocracy first articulated in *The Art-Journal* in 1855. The term "technostress" was coined by industrial psychologist Craig Brod in 1982, in the title of his paper *Managing Technostress: Optimizing the Use of Computer Technology*. The article abstract is ominous: "Afraid of change, many employees refuse to join the computer revolution. The author states that only understanding and proper training can break down such resistance." Brod defined technostress as a "modern disease of adaptation caused by an inability to cope with new computer technologies in a healthy manner."[50] By problematizing people's reaction to increasingly pervasive technology in all facets of daily life, rather than questioning implications of the increasing pervasiveness of technology, Brod was clearly working in the tradition of Scientific Management first conceived by Frederick Winslow Taylor at the turn of the twentieth century. Writing for the management-oriented periodical *Personnel Journal*, Brod was interested in how to use psychology to impose new workplace technologies on workers through hierarchical chain of command. Rather than problematizing the omnipresence of technology, Brod problematized the worker who did not adapt to the workplace environment.

As the workplace became more technocratic in the late twentieth and early twenty-first centuries, so too did modern war work. The comprehensive nature of war brought about revised notions of shell shock, which considered more nuanced forms of trauma than Charles Myers's initial diagnosis. The term Post Traumatic Stress Disorder (PTSD) has its origins in the American antiwar movement of the 1960s. Dr. Chaim Shatan's experience with Vietnam Veterans Against the War prompted him to write on op-ed piece in the *New York Times* on May 6, 1972 titled "Post-Vietnam Syndrome." In his article, Shatan observed that feelings of guilt, scapegoating, rage, brutalization, alienation, and doubt result in "'impacted grief' in which an encapsulated, never-ending past deprives the present of meaning."[51] From this initial observation and subsequent psychological inquiries, the term "Post Traumatic Stress Disorder" was accepted by the American Psychiatric Association into the *Diagnostic and Statistical Manual of Mental Disorders* (the DSM-III) in 1980. It listed as the first diagnostic criteria for identifying PTSD as follows:

> The person has experienced an event that is outside the range of usual human experience and that would be markedly distressing to almost anyone, e.g., serious threat to one's life or physical integrity; serious threat or harm to one's children, spouse, or other close relatives and friend; sudden destruction of one's home or community; or seeing another person who has recently been, or is being, seriously injured or killed as a result of an accidental or physical violence.[52]

The extent to which these circumstances, borne of martial violence and warfare, are prevalent in civilian life is evidenced by the exponential number of PTSD cases diagnosed since it became a viable medical opinion in 1980. While the DSM-III definition was based on the military's diagnostic criteria, the DSM-IV and V contain substantial revisions to account for the degree of conditions and symptoms that can be diagnosed as PTSD, accounting for, among other things, the prevalence of PTSD in women and children.[53] In the United States, 7.8% of the adult civilian population has experienced PTSD at some point in their lives, with women twice as likely as men to develop symptoms.[54] Not only modern war, but modern life itself, had become stressful to the point of affliction.

Etymologically then, technocracy has dual meanings. It is simultaneously a term of liberation and oppression. *The Art-Journal's* use of the term implied a cultural assault on the senses, where William Henry Smyth's use of the term suggested political and economic liberation. Over time, with the onset and spread of modernity, the oppressive qualities of technocracy with which the original use of the term was infused have won out. Technocratic modernity is traumatizing. Like the working classes who were left to suffer its conditions at the onset of the Industrial Revolution, those who cannot afford to escape its omnipresence are constantly subjected to it. Modernity wakes us in the morning, and exhaustion sets us to sleep at night. Our waking hours, governed by technocracy, are lived within the bounds of the monoculture which is as comprehensive as it is pervasive.

MONOCULTURE

Like technocracy, the term monoculture has dual meanings. The first is agricultural, the second is social. The two meanings are connected, and share the same roots of absolutism, slavery, capitalism, and imperialism. To understand their connection, we should work backwards, considering the social meanings of monoculture first, and then putting those understandings into the agricultural context from which those social circumstances originate.

The first historians to connect the agricultural practice of monoculture to the sociopolitical conditions of early modernity were Perry Anderson and Richard B. Sheridan. In 1974, both wrote books which identified tentative but important links between single-crop agronomy and the cultures in which they developed. In *Lineages of the Absolutist State*, Anderson observed that in seventeenth-century Poland, "an agrarian monoculture was thus increasingly created, which imported its manufactured goods from the West in an aristocratic preconfiguration of the overseas economies of the nineteenth century." He went on to conclude that "the noble class which emerged on these economic foundations had no exact parallel anywhere else in Europe." Here, the imposition of monocultural agricultural labor on the peasantry left them destitute and powerless. Unable to provide for themselves through subsistence farming or in a wage economy, Anderson observed that the peasants were subjected to "the 'colonial' character of the landlord class" while the nobility accumulated unprecedented amounts of wealth and power.[55] His conclusion linked the accumulation of wealth that agricultural monoculturalism provided with the accumulation of power that the sociopolitical conditions of agricultural monoculturalism created:

> The new and singular type of *State* that arose in this epoch was Absolutism. . . . The dominant state structure in Europe down to the end of the Enlightenment, its ascendancy coincided with the exploration of the globe by European powers, and the beginning of their supremacy over it. In nature and structure, the Absolute monarchies of Europe were still feudal states: the machinery of rule of the same aristocratic class that had dominated the Middle Ages. But in Western Europe where they were born, the *social formations* which they governed were a complex combination of *feudal and capitalist modes of production*, with a gradually rising urban bourgeoisie and a growing primitive accumulation of capital, on an international scale. It was the intertwining of these two antagonistic modes of production within single societies that gave rise to transitional forms of Absolutism.[56]

Also in 1974, Richard B. Sheridan's work, *Sugar and Slavery: An Economic History of the British West Indies, 1623–1775*, made explicit connections between agricultural monoculture, slavery, capitalist monopoly, and the consolidation of imperial power. "The near-monoculture sugar regime" as

Sheridan put it, "was dominated by a class of wealthy sugar planters." Its first manifestation on Barbados "was so disruptive that thousands of smallholders and indentured servants were forced to seek their fortunes elsewhere."[57] In Jamaica, like in Anderson's observations of Poland, "the plantation economy was tending toward monoculture and taking away the livelihood of smallholders and artisans." Here, agricultural monoculture's social and political implications were obvious. The diversity of the land and the people who lived on it were erased to make way for hierarchical, uniform control for the sole purpose of capital accumulation. Sheridan's use of sources made plain the connection between soil exhaustion and labor exhaustion. This citation from John Oldmixon's 1708 *The British Empire in America* is unambiguous:

> The planters being limited to a small Proportion of land, pressed it so often with the same Plant, and never letting it lie still, the Soil is so impoverish'd, that they are now forc'd to dung and plant every Year; insomuch that 100 acres of Cane require almost double that Number of Hands they did formerly . . .[58]

The biological exhaustion produced by monoculture is comprehensive. To this point, Sheridan observed that "in the development of a given colony, opportunities for escape and independent sustenance were probably greatest in the infancy of the sugar industry, while violent resistance tended to mount during the transition to monoculture. Thereafter, resistance and flight continued, to be sure, but probably at a reduced tempo."[59]

In this context, the early modern form of monoculture, rooted in large-scale agricultural production, rendered the earth and the people who lived on it subject to the consolidation of political and economic power. It wasn't until 1979 that Gary Snyder, the American poet and environmental activist associated with the Deep Ecology movement, connected the politically and economically comprehensive nature of monoculture with the social conditions that exist within it:

> There is a huge investment in this nation. . . . All that belongs to somebody, and they don't want to see it become useless, unprofitable, obsolete. In strict terms of cash flow and energy flow it still works, but the hidden costs are enormous and those who pay the costs are not the owners. Their investment requires continual growth, or it falters. . . . I repeat this well known information to remind us, then, that monoculture-heavy industry, television, automobile culture—is not an ongoing accident; it is deliberately fostered. Any remnant city neighbourhood of good cheer and old friendships, or farming community that 'wants to stay the way it is' are threats to the investment.[60]

With this, monoculture becomes not just a means of production based on political and economic control, but a means and an ends of social conditioning, which constantly reinforces itself through pervasive expansion. At its base, monoculture depends on the enforcement of assimilation to take root and expand. The enforcement of assimilation comes from the hierarchically modeled political, economic, and social structures that use the tool of monoculture as a means of regulating and ordering space and the people who occupy it. Monoculture, originally a physical imposition, has become an ideological imposition within the framework of modernity. Its comprehensive nature is the product of modernity itself. Monoculture is fostered by technology, the dual forces of nationalism and urbanization, and the market economy. Clearly at the expense of a democratic ideology, it seeks to either commodify or destroy altogether.

Technology enables the territorial and ideological expansion of monoculture through its pervasive communication and transportation networks. Here, the phenomenon of time-space compression as articulated by David Harvey in his 1989 book, *The Condition of Postmodernity*, and the phenomenon of time-space distanciation as articulated by Anthony Giddens a year later in *The Consequences of Modernity* both inform the ideological expansion of monoculture. In Harvey's definition of time-space compression, "a strong case can be made that the history of capitalism has been characterized by a speed-up in the pace of life, while so overcoming spatial barriers that the world sometimes seems to collapse inwards upon us."[61] Innovations in communications and transportation, in Harvey's view, "annihilate space through time." In Giddens' consideration of time-space distanciation, modern technology enables the projection of power and authority in the absence of any physical presence. This allows for institutional authority to be "disembedded" from time-space, and their rational organization of hierarchical political, economic, and cultural models becomes ubiquitous.[62] Both of these phenomena, rooted in technological modernity, ultimately serve to project the conditions of capitalistic monoculture onto the population within its ever-expanding reach.

The very idea of globalization, driven exclusively by market capitalism, promotes a universality of monocultural homogeneity that envisions people unified, not by ideas or ideals, but by markets. As Jonathan Friedman observed in his 1990 essay on the dual phenomena of globalization and localization: "The dualist, centralized world of the double East-West hegemony is fragmenting, politically and culturally, but the homogeneity of capitalism remains as intact and as systematic as ever."[63] The expansion of markets is the single most important aspect of capitalism. Economic globalization is not cooperative, it is competitive. Barbara Bradby's assertion that "the establishment of capitalism in a social formation necessarily implies the transformation, and in some sense, the destruction, of formerly dominant modes of production" informs the terms and conditions of globalization.[64] In a sense, capitalism is itself monocultural. In destroying the heterogeneity of political, economic, and cultural conditions it encounters, it lays bare the soil for the establishment of its own rules of order: hierarchical, singular, and comprehensive.

The tactic of profiting from the crises that occur in the wake of destruction is a phenomenon that Naomi Klein elaborated on in her 2007 book, *The Shock Doctrine*. Klein's articulation of "disaster capitalism," a term she uses to describe the "orchestrated raids on the public sphere in the wake of catastrophic events, combined with the treatment of disasters as exciting market opportunities," is, again in her words, "a challenge to the central and most cherished claim in the official story—that the triumph of deregulated capitalism has been born of freedom, that unfettered markets go hand in hand with democracy." Klein asserts that "fundamentalist form of capitalism has consistently been midwifed by the most brutal forms of coercion, inflicted on the collective body politic as well as countless individual bodies. The history of the contemporary free market—better understood as the rise of corporatism—was written in shocks."[65]

The agricultural roots of monoculture are ancient, and resonate loudly in this context. Karl Wittfogel considered the original connections between agriculture and power in his 1957 monograph, *Oriental Despotism: A Comparative Study of Total Power*. Wittfogel argued that the origin of power in ancient societies was bound up with the hierarchical control of water access. He used the term "hydraulic civilization," and used Marx's concept of the "Asiatic Mode of Production" to assert that "Oriental Despotism" was "more comprehensive and oppressive than its Western counterpart." Based in his readings of sixteenth- and seventeenth-century economists like Bernier and Montesquieu, "Oriental Despotism presented the harshest form of total power."[66] What Wittfogel recognized was that the modern bureaucratic state has its beginnings in the framings of power that centered around control of water for purposes of agricultural production. What Wittfogel got wrong was that the "Asiatic Mode" was somehow an isolated design which came to infect the western world, leading to a corruption of Marxist teleology. Wittfogel's ultimate purpose was to show that Soviet agricultural policy in the 1950s was more "Oriental" than Marxist, and this was the reason for its failure. Polemical reasoning aside, Wittfogel's idea of "hydraulic civilization" as hierarchical and absolutist is informative, and the despotism which results is universally recognizable.

Taking nothing away from the despotic elements of hydraulic civilizations in ancient Sumer, Egypt, China, and India, evidence of despotism based on the control of agricultural resources, including water, are evident in ancient Rome as well. Unfortunately, systematic oppression in some form or another seems to be a relatively universal historical quality. The institution of *latifundia* (Latin for "spacious land"), vast estates owned by the patrician classes who used their influence for to compound political and economic gain, were one of the many causes of the fall of the Roman Republic and the rise of the Roman Empire in the first century BC. Conquered lands were spoils of war for the Roman army, and soldiers who could not compete on the fast consolidating land market sold to larger estates, typically owned by senators who did not pay land taxes. This consolidation led to the territorial growth of the *latifundia*; as the Roman Empire expanded, so too did the profits of the principal landholders who directly shaped the imperial war policies.

The consolidation of power and wealth in ancient Rome, which led to the death of its representative form of government and the ascent of imperial authority, was a result of agricultural appropriation and control. Christopher Francese suggests that the word *latifundia* "means essentially 'plantation' but it has a bit of the flavour of the English compound 'agribusiness,' in that it symbolizes the decline of the idyllic family farm and its replacement by soulless industrial farming."[67] Pliny the Elder lamented in his work *Natural History*, written during the reign of the emperor Titus in AD 79, that "if we must confess the truth, it is the wide-spread domains that have been the ruin of Italy, and soon will be the ruin of the provinces as well."[68]

Slavery was essential to the consolidation of land-holdings by the Senatorial class in Ancient Rome. Pliny observed that the conditions of slavery on the *latifundia* were antithetical to the virtue of Republican identity: "It is the very worst plan of all, to have land tilled by slaves let loose from the houses of correction, as indeed, is the case with all work entrusted to men who live without hope."[69] When the cost of slaves increased toward the end of the Roman empire, tenant farmers worked the land and laid the transitional foundation for the feudal manorial economies of medieval Europe.[70] The *latifundia* system returned as early modern Europe began the processes of enclosures, transatlantic territorial expansion and industrialization in the sixteenth through the eighteenth centuries. Slavery and subjection followed in the wake of the agricultural and industrial factory systems. Slave revolts in the Americas and worker revolts in Europe pushed back against the despotism of monoculture.

When the United States gained independence from British Empire, its government looked west to develop the land as its primary source of wealth. Key to this development, as Donald Worster pointed out in his 1985 work, *Rivers of Empire*, was the implementation of "The Capitalist State Mode" of production, where two centers of power, one public and one private, both reinforce each other and compete against each other to "achieve a control over nature that is unprecedentedly thorough."[71] Worster observed that the organization of water resources in the western United States was elementally consistent with Karl Wittfogel's idea of "hydraulic despotism," and in some ways, surpassed the comprehensive nature of Wittfogel's "Oriental" construct. He explained that:

> The most fundamental characteristic of the latest irrigation mode is its behaviour toward nature and the underlying attitudes on which it is based. Water in the capitalist state has no intrinsic value, no integrity that must be respected. Water is no longer valued as a divinely appointed means for survival, for producing and reproducing human life, as it was in local subsistence communities. Nor is water an awe-inspiring, animistic ally in the quest for political empire, as it was in the

agrarian states. It has now become a commodity that is bought and sold and used to make other commodities that can be bought and sold and carried to the marketplace. It is, in other words, purely and abstractly a commercial instrument.[72]

The agricultural form of monoculture carries with it the same implications of the social form of monoculture. Capitalistic commodification and control, aided by technology, establishes a uniformity of conditions for the benefit of a few at the expense of the many. Within the conditions of modernity, agricultural and social monoculture are woven together to produce a more comprehensive result.

Corporate control of the food supply in the modern world produces both agricultural and social monoculturalism. Both render the body biological and the body politic passive, lethargic, and compliant. Agribusiness has become a monolithic presence on the American landscape, and throughout the world. Corporations like Archer Daniels Midland, Cargill, Monsanto, Perdue, Smithfield, and Tyson dominate the corporate landscape of food production, operating with relative impunity in "The Capitalist State Mode" of production, consolidating wealth and political power at the expense of the population subject to the terms and conditions they are able to impose, both through legislative lobbying and by monopolizing the marketplace. While supermarkets offer a superficial veneer of choices to consumers, most of their food comes from a relatively few sources, who spend hundreds of millions every year on corporate lobbying and legal actions to protect and expand their market shares.[73]

The term "monoculture" itself is a product of early twentieth-century industrial agriculture. Changes in farming technology, particularly the availability of chemical fertilizer and gas powered tractors, signaled fundamental changes in American agriculture at a time when industrialization was giving rise to urban populations. These technological developments made supply soar while demand remained level, forcing farmers to mortgage their land to buy more of these technological "necessities." The result, as James Howard Kunstler observes in his seminal work *The Geography of Nowhere*, is the death of agriculture and the rise of agribusiness, and the industrial production of food.[74] The tractor and chemical fertilizer together made it possible for farmers who had the capital to afford them to work exponentially large tracts of land and to specialize in one crop without sacrificing yield. Smaller, freehold farmers, without the capital necessary to compete, either sold or lost their farms, and moved to the new industrial centers to find work.

In both social monoculturalism and agricultural monoculture, the interests of market economics and technology undermine democratic ideology. Over historical time, across geographical space, the spread of monocultural hegemony, first in national terms, then in the globalized market, represents a fundamental challenge to teleological progress. The comprehensive nature of monoculture makes it difficult to articulate, much less to comprehend.

We eat monoculture, we live in monoculture, and we often unconsciously reinforce monoculture through our words, actions, and thoughts. And so monoculture both eats itself and feeds itself: a modern Oroboros, which Plato described in *Timeaus*:

> The living being had no need of eyes because there was nothing outside of him to be seen; nor of ears because there was nothing to be heard; and there was no surrounding atmosphere to be breathed; nor would there have been any use of organs by the help of which he might receive his food or get rid of what he had already digested, since there was nothing which went from him or came into him: for there was nothing beside him. Of design he created thus; his own waste providing his own food, and all that he did or suffered taking place in and by himself.

In a passage that suggests the ontological stasis which technocratic monoculture imposes, Plato further observed: "and he was made to move in the same manner and on the same spot, within his own limits revolving in a circle. All the other six motions were taken away from him, and he was made not to partake of their deviations."[75] Monoculture goes nowhere.

THE TERRITORIAL IMPOSITION OF TECHNOCRATIC MONOCULTURE

Technocratic monoculture develops in space, over time. The triangulation of both for the purpose of hegemonic authority and control was begun in earnest in 1675, when King Charles II commissioned the construction of the Royal Greenwich Observatory "in order to the finding out of the longitude of places for perfecting navigation and astronomy."[76] The concept of longitude was articulated by the Greek geographer Eratosthenes in the third century BC, but it was not until the fifteenth century that competition among the emerging global powers for the technical capability to determine precise longitude became a driving factor of scientific inquiry. With the development and expansion of transatlantic imperial economies, the necessity for geographical measurement to impose control and authority over territories and shipping lanes became vital to the expansion and mutual reinforcement of state and corporate control of space. Thus, the origins of modern timespace are particularly linked to the development of the systems that define the epoch as something distinct from the medieval and ancient. At Greenwich in 1675, we find the convergence of technology and capital in the pursuit of state-sanctioned imperial wealth and power. The triangulation of vague borderlands, and the imposition of sharply defined and enforced borderlines, is also an essential component in the manifestation and administration of technocratic monoculture.

The connections between space and time at Greenwich are intertwined and inextricable. From the astronomical observations conducted at Greenwich, the development of spatially accurate cartographic models based on the conception of longitude was predicated on a precise knowledge of time. In pursuing the triangulation of space for the pursuit of imperial objectives, technological advances for the purpose of measuring time became just as important to the control of imperial space, resulting in a comprehensive hegemony of order that has become exponentially expansive ever since. Greenwich, therefore, is the center of the temporal and spatial framework for the modern world. The global imposition of the systems designed at Greenwich in the seventeenth century represent the imperial ambition of universal, ontological ordering, which, when connected to the principle of sovereign hegemony, has served as justification for the exercise of unchecked hierarchical authority.

The English imperial design held sway over much of the planet in 1763, with the signing of the Treaty of Paris ending the Seven Years War. In the aftermath of the American Revolution which followed, the United States looked to expand its political control over territory west of the Appalachian Mountains. The United States of America picked up the expansive nature of temporal and spatial imperialism when they won their independence from Britain in the American Revolution. The newly formed U.S. government was heavily indebted, and turned its attention to developing its primary source of wealth: the land. Here, spatial triangulation for the purpose of commodification was systemized on what the United States Army enforced as a tabula rasa, pushing native populations westward to the periphery in a concerted effort to make room for land speculators and developers, who would turn the land economically productive, expanding the market economy over space, and instituting it over time.[77] "From the point of beginning," the Land Ordinance of 1785 began establishing a rectilinear grid over the North American continent from the Ohio River westward.

Thomas Jefferson was particularly interested in filling in the "Terrae Incognitae" of his expansive map collection, sending a range of explorers, naturalists, and wanderers out to collect information which would make political, economic, and cultural control of the space west of the Appalachian Mountains more complete. In 1751, his father Peter Jefferson, along with Joshua Fry, had completed the most technically accurate map of Virginia that had been made to date (See Figure 3.2). Peter Jefferson understood that spatial triangulation was key to control and appropriation of territory, and with it the people who already inhabited it along with the resources they relied on to live. In the *"Map of the Most Inhabited Part of Virginia . . ."* the western limits of the colonies are greatly expanded, and the representation of native populations greatly reduced. The names of the English colonies are superimposed with imperial authority over rivers, trees, and mountain ranges on the Jefferson-Fry map. The recognition of Indian presence is greatly reduced. Indians are literally marginalized on the

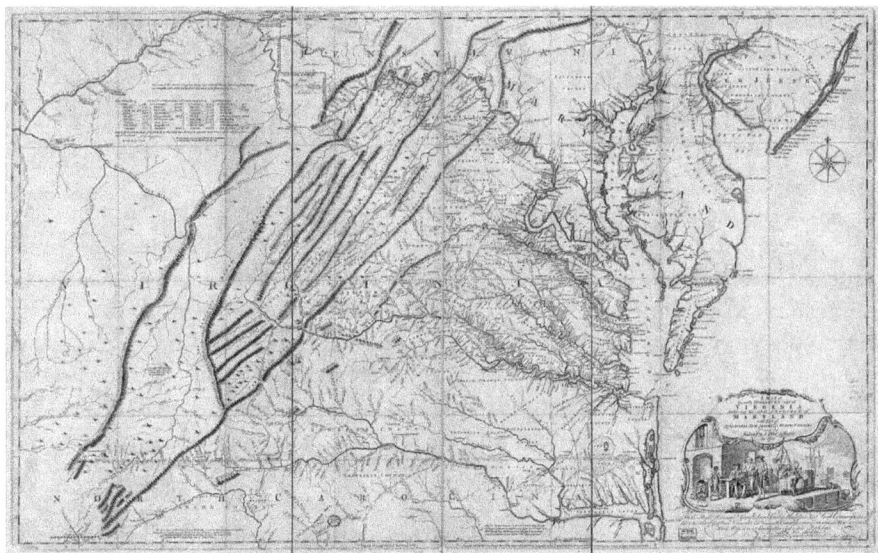

Figure 3.2 A map of the most inhabited part of Virginia containing the whole province of Maryland with part of Pensilvania, New Jersey and North Carolina.
Drawn by Joshua Fry & Peter Jefferson in 1751. Library of Congress, Geography and Map Division. http://hdl.loc.gov/loc.gmd/g3880.ct000370.

map; they are removed from the bounded articulation of the colonies and relegated to lower case lettering for representation. The Jefferson-Fry map was primarily designed to encourage English investment in British North America by depicting an essentially "safe" investment. The message of easy money was amplified by the overtly racial map cartouche depicting African slaves moving large quantities of goods for transatlantic shipment while Anglo merchants oversaw their productivity and enjoyed refreshments being offered by yet another African slave.[78]

Filling the cartographic void that existed between the Mississippi River and the Pacific Ocean became an obsession that Thomas Jefferson inherited from his father, and the urgency of his mission intensified after he had arranged for the "purchase" of the Louisiana Territory from a cash-strapped Napoleon in 1803. Realizing that survey knowledge of the territory was necessary for the political and economic control, Jefferson hastily organized an official expedition to take measure of the land before the treaty with France had even been ratified by the Congress. "The object of your mission," Jefferson wrote to Meriweather Lewis, whom he had appointed to lead the expedition, "is to explore the Missouri river, & such principal stream of it as by it's [sic] course and communication with the waters of the Pacific ocean whether the Columbia, Oregon, Colorado or any other river may offer the most direct & practicable water communication across this continent for the purposes of commerce."⁷⁹ Commercial interests drove the territorial expansion of American nationalism at the turn of the nineteenth century: a sort of continental imperialism that imposed terms on the indigenous populations, which would ultimately lead to their internment onto reservations, depopulation, and ultimate exclusion from the system of commerce as it spread westward.

And commerce spread westward quickly. As ambitious plans go, the pursuit of territorial hegemony went spectacularly. By the end of the nineteenth century, the indigenous peoples of North America had been reduced to a mere shadow of their former numbers, relegated to reservations which institutionalized poverty and alcoholism from which the present day native populations are just beginning to recover, albeit haltingly and incongruously. While Jefferson had expected that it would take hundreds of generations for Americans to populate the territory west of the Appalachian Mountains, it would take only ninety years from the date of the Louisiana Purchase for the historian Frederick Jackson Turner to open his seminal paper, "The Significance of the Frontier in American History" with these lines:

> In a recent bulletin of the Superintendent of the Census for 1890 appear these significant words: 'Up to and including 1880 the country had a frontier of settlement, but at present the unsettled area has been so broken into by isolated bodies of settlement that there can hardly be said to be a frontier line. In the discussion of its extent, its westward movement, etc., it can not, therefore, any longer have a place in the census reports.' This brief official statement marks the closing of a great historic movement. Up to our own day American history has been in a large degree the history of the colonization of the Great West. The existence of an area of free land, its continuous recession, and the advance of American settlement westward, explain American development.⁸⁰

Over that course of time, the triangulation of the American continent was nearly completed. The rough grid laid out by Gunter's chain and the theodolite is clearly visible from above, but at ground level, it represented the initial abstraction of the land that would allow speculative investors to control it from remote locations, to claim possession of its raw materials and to evict or charge rents to its occupants accordingly. The enforcement of this remote control, established in the nineteenth century, would become a central theme of the twentieth century, as methods of surveillance to monitor, police, and regulate the comings and goings of labor (workers) and capital (raw materials and products) developed. Control of space, and the activity occurring within those spaces, became a primary objective of the nascent technocratic monoculture. The result of such development was the enforcement of capitalist order. Squatters were removed from the land, infrastructure was superimposed on the landscape through the use of eminent domain, which allowed corporations to take land from freeholders in order to implement their industrial designs; displaced citizens moved to cities as industrial capital became the ascendant form of economic exchange.

Controlling space and the people who lived in it had been a principle objective of the architects of this neo-Olympian order as Chaos shook the peripheral frontier. The threat came from four groups: indigenous populations, slaves, small freehold farmers and itinerant workers, and combinations thereof. They remained outside the bounds of law, both spatially and ideologically, like the many beasts just beyond the bounds of the sovereign's power. The threat of Kronos loomed; liberty was on everyone's lips, and all Olympus could do was to unchain the Cyclopes to meet the threat of Chaos. To do so, laws designed to incarcerate or marginalize resistance and economic conditions intended to render free people dependent and subject were imposed from the top down, imposing an order that was hierarchical and self-sustaining. Technocratic monoculture moved from the triangulation of borderlands to borderlines, and those borderlines then shifted from territorial to ideological control.

SECURITY VS. FREEDOM

A significant feature of technocratic monoculture's ability to police both territorial and ideological space is the promotion of security at the expense of freedom. Security is an inherently political and economic condition, not to be confused with the social condition of safety. Security is based on suspicion, while safety is built on trust and mutuality. The growth of the security economy in the twentieth century is inherently a function of the market economy, selling freedom from fear as a product for consumption. The basis of the security economy is the maintenance and protection of private property. As the amount of private property grew exponentially as a result of transatlantic industrialization in the nineteenth century, the need to protect the accumulation of wealth became a central concern of the newly propertied classes. Its origins lie in the early modern appropriation of material wealth which created a hierarchy of economic disparity, prompting the need for increased security measures to protect the exponential accumulation that resulted.

The rise of the security market that accompanied this increase in the disparity of material wealth prompted George Price of Wolverhampton to write *A Treatise on Fire & Thief Proof Depositories and Locks and Keys* in 1856. Price attributed the origins of the Iron Safe to the late seventeenth and early eighteenth centuries, citing that "our forefathers, in the simplicity of their arrangements and requirements, were satisfied to place their valuables in an oak chest," and that "the first examples of the manufacture of iron safes, or chests, are the foreign coffers." These crafted prototypes came from France and Germany, introduced to England in the early eighteenth century, and "must have been considered curiosities." When the iron safe was applied "for commercial purposes," its manufacture was industrialized "at Coalbrook-dale, Birmingham, Wolverhampton, and at some other places, and have been exported to all parts of the world."[81] Price goes on to warn his reader about the dangers of storing their valuables in a safe of substandard design, deigning them "utterly worthless as preservers of their contents against fire or thieves" because they "readily yield to the blow of a hammer, or may be easily opened by the use of the drill."[82] After selling his reader the fear of having an inferior safe, Price offers a solution: *Price's Patent First Class Fire Resisting and Thief Proof Safe*, ensuring potential customers that "these safes are well adapted for the preservation of cash and other property of great value from burglars . . . also for the preservation of books, banknotes &c., from damage by fire."[83]

George Price cites Granville Sharp's "Prize Essay on Practical Banking" as an influential source of his work. Not to be confused with the English abolitionist who shared the same name, this Granville Sharp submitted his essay in 1851, assuring his readers that "considerable pains have been taken to render it serviceable to Bankers, and to Bankers' Clerks." One of the principle themes of the essay was "new inventions in the construction of locks, cash boxes and safes, which shall render property more secure against fire or thieves." Here Sharp offers background to the research and design that goes in to making institutions of commerce more secure:

> The protection and preservation of life and property, are objects which have been anxiously and necessarily attended to in the simplest states of society; and the invention of ingenious men has in all times been applied to contrive means of security adapted to resist the nefarious practices of the day. A review of their productions, which we must presume to have been effectual to their purpose, suggests a conclusion, that the morals of former times were as much less depraved than those of the present, as the various mechanical contrivances were less excellent; and from the various methods of protection which have been successively used, it may be collected that the arts of violation have improved at least in an equal degree. It is certain that no invention for the security of person and property has yet been offered to the world, which the ingenuity of wickedness did not find means to defeat.

Sharp concludes with an ominous warning: "Modern depredation is reduced to a system, in which art and force are exerted with such skill and power, as to elude precaution, and to defy resistance."[84] This is the formative language of security. Human nature is inherently depraved, and grows more so with each passing moment. Protection from it depends on investment in a security apparatus, which will become obsolete when its technological walls are breached by "the ingenuity of wickedness."[85] Security is an economic condition; its principle product is fear.

Since the expansion of material wealth that came with the technological developments of the Industrial Revolution, the expansion of the security market has grown accordingly. According to the 2013 ASIS Institute of Financial Management Report, "The United States Security Industry: Size and Scope, Insights, Trends and Data," the private security industry in the United States is a $282 billion dollar industry which employs between 1.9 and 2.1 million people.[86] This exponential expansion of the security industry is entirely a product of modernity: the advance of technology to protect the interests of capitalist market economics. By advancing the control of space to protect private property, the accumulation of wealth continues at the expense of the larger numbers who suffer as a result of the disparity.

The growth of the security industry in the United States is inexorably linked to the increased disparity of wealth in the United States, which has become institutional over the course of the nineteenth and twentieth centuries. This process has accelerated since 1980 with the introduction of "Reaganomics," a collection of economic policies that ostensibly freed the wealthy from any binding financial obligation to the society from which they extract their wealth. Since then, the richest 10% of Americans possess two-thirds of the nation's wealth, and the top 1% has seen their share of annual income grow 125% while the share of annual income for the bottom 80% of America's wage earners has dropped from 5% to 30%.[87] Given this unprecedented consolidation of wealth, the growth of the private security sector based on fear seems a foregone conclusion.

The spatial expansion of security as a means of protecting wealth has grown over this same period. Edward J. Blakely and Mary Gail Snyder trace the history of gated communities in their book *Fortress America: Gated Communities in the United States*, locating the origins of gated communities within the historical framework of western civilization, ancient Rome, medieval England, and early colonial America. They note that the proliferation of gated communities in the 1980s "marked the emergence of gated communities built primarily out of fear, as the public became increasingly preoccupied with violent crime."[88] The spatial apparatus of control is explicit:

> Economic and social segregation are not new. In fact, zoning and city planning were designed, in part, to preserve the position of the privileged with subtle variances in building and density codes. But gated communities go further in several respects than other means of exclusion. They create physical barriers to access. They also privatize community space, not merely individual space.[89]

They go on to observe that "the first step to creating this private world is controlling access to it," and that "over time, developers have devised many means of controlling access."[90]

Those many means are something that the urban planner Daniel D'Oca calls "The Arsenal of Exclusion & Inclusion." This is an increasingly sophisticated set of codes and symbols embedded in planned physical space, which represent a battlefield on which the agents of technocratic monoculture and those who seek to undo it "wage the ongoing war between integration and segregation, between NIMBY (not in my backyard) and WIMBY (welcome in my backyard)." D'Oca's comprehensive assessment of urban space identifies weapons in this arsenal like armrests and one way streets. Seemingly innocuous, D'Oca observes that armrests are use: "to deter the homeless from sleeping on park benches, decorative armrests are sometimes installed at the midpoint of the benches, making it impossible (or at least very difficult) to get too comfortable on them, and that one way streets are used to disconnect less affluent neighborhoods from wealthy neighborhoods by preventing the flow of traffic from the former to the latter.[91]

The arsenal of exclusion in this present form has origins in earlier versions of spatial exclusion, most noticeably in the enclosure movement of early modern England, which systematically appropriated common land as private property. Enclosure was a long process, with roots in medieval manorial arrangements, but from the seventeenth to the nineteenth century, a critical mass of legislation against common rights to grazing, pasturage, and farming created a small, propertied class at the expense of a newly landless working class, who were denied their traditional rights to the land and forced to find wage work where they could.[92] In 1684, Thomas More wrote of the effect that the process of enclosure had on tenancy:

> for when any unsatiable wretch, who is a Plague to his Country, resolves to inclose many thousand acres of ground, the Owners, as well as Tenants, are turned out of their possessions, by Tricks or by main Force. . . . So those miserable People, both Men and Women, Married, Unmarried, Old and Young, with their Poor but numerous Families (since Country-Business requires many hands), are forced to change their Seats. . . . What is left for them to do, but either to steal, and so to be hanged, (God knows how justly), or to go about and beg? And if they do this, they are put in Prison as idle Vagabonds . . .[93]

Historians have pointed to the enclosure system as a primary factor in modern urbanization, as the newly landless began populating cities to find work in the burgeoning field of manufactures. Enclosures were enforced both spatially and legally. Fences established a perimeter to protect newly privatized property, and deeds protected an owner's rights to exclusivity within those bounds.

Laws to impose and enforce work discipline in the new wage economy were passed alongside enclosure bills, creating a legal framework with which to enforce the proletarianization of peasants and commoners in a new, modern, industrial economy. The deracinated population soon found strict laws aimed against the poor, including laws against vagabondage. Those found guilty under Henry VIII at the turn of the sixteenth century were subject to whipping, mutilation, and hanging. Midcentury, Edward VI increased the penalties to include branding and terms of slavery. By the end of the century, Elizabeth I saw that vagabonds were whipped and banished to galley service.[94] As Marx put it: "Thus the agricultural people, [were] first forcibly expropriated from the soil, driven from their homes, turned into vagabonds, and then whipped, branded, tortured by laws grotesquely terrible, into the discipline necessary for the wage system."[95]

This systematic eviction of the peasantry from common pasturage and tillage, and the legislation that criminalized their traveling resulted in a general suspicion of the dispossessed. They represented a chaotic element within a new order, beasts in the landscape of the sovereign. Moral framings which typecast vagrants lacking of work as criminal, debauched, and feckless were enforced in the courts as well as the institutions of learning and correction. The "unproductive independence" of the vagabond was a threat to the emergent social, political, and economic order, which sought to impose work discipline, substandard wages and living conditions, and disposability on the newly formed working class. The growth of the range of property in physical and in legal terms led to the increased marginalization and peripheralization of the homeless population. Vagrancy acts continued to be passed throughout the nineteenth and twentieth centuries, on both sides of the Atlantic, as new methods of control aimed at incarceration and correction rather than physical abuse and expulsion became standard. Still, the typecast of the vagabond, vagrant, hobo, or homeless person remained one of "unproductive independence," ideologically problematic in the context of an expanding technocratic monoculture that is systematically designed to impose the condition of "productive dependence" upon its subject population.

The Reverend Frank Laubach concluded in his 1916 study, *Why There Are Vagrants*, that "without doubt, many men begin the wandering life in revolt against the monotony of modern industry. It can hardly be expected that the descendants of the most daring and adventurous of the nations of the earth should all be willing to submit to humdrum, unchanging toil."[96] Laubach further observed that "the rich man may travel all he will, but the poor man . . . must be satisfied with a limited horizon, unless he 'tramps it,' or rides the freight."[97] The wealthy exercised spatial liberation; the poor were incarcerated. The mobility of the poor within the technocratic monoculture was something to be problematized, policed, and prevented. In cities, an increasingly expansive arsenal of exclusion was directed at channeling the homeless into spatially designated areas for the purposes of command and control, Such as philanthropically run shelters or skid rows, where they could be reformed or policed accordingly. Outside of city limits, on the periphery beyond the reach of the sovereign, hobos were less easily policed, monitored, and controlled.

The construction of homeless shelters in large cities was part of a movement at the turn of the nineteenth century to address the subject of homelessness through modern, bureaucratically organized philanthropic and civic organizations. In shelters, the homeless were expected to work in exchange for temporary shelter. For the Charitable Organization Society, which operated shelters in Boston, Cincinnati, New York, and Chicago, among other cities, in the late 1800s, the goal was to instill a work ethic in the homeless population through a "mixture of force and kindness."[98] For the homeless urban populations who avoided the lodging houses run by charity organizations, skid rows situated in disregarded urban areas were contained by professional police forces, which reinforced class boundaries by keeping the homeless confined and marginalized within the city limits.[99]

Outside of the cities however, the homeless population had more autonomy. On the peripheral outskirts of cities and towns, "hobo jungles" provided transients with resources, information, and community. Here was a revival of the commons, autonomous and collective. In his 1930 book *The Milk and Honey Route: A Handbook for Hobos*, the sociologist Nels Anderson wrote of the collective nature of resources in the hobo jungles: "Here you share and share alike in true fraternal style."[100] These improvised communities, multiethnic and multiracial, threatened the infrastructure of hierarchical control by reinventing the preindustrial values of the commons, and as such were outside of the law and order of the day. They were below and beyond, "on the bottom and on the fringe" as the sociologist Peter H. Rossi puts it, and as such were subject to the institutional rule of law which criminalized their existence and legitimized their destruction. Hobo jungles, quasi-permanent arrangements that offered an alternative to modern arrangements of fixed labor and living, played cat and mouse with authorities throughout the nineteenth and twentieth centuries.

Railroad workers often acted indiscriminately in defending their employer's right to exclusivity, and retained a degree of legal protection for their abuses. As the historian Kenneth Kusmer notes: "A carrier owed no duty to a trespasser on a train and could eject him at any time. If the person was injured in the process, he could not recover damages except in the rare instance that the train crew had attempted to 'wantonly' or 'recklessly' injure him." Quoting an article in *The Railroad Gazette* from 1883, Kusmer throws light on the understandings between the railroad workers and law enforcement at the time: "If a brakeman throws a tramp off a train and he is killed, you will generally read an item about an unknown tramp, while trying to steal a ride, having fallen between the wheels or something of that kind, but we know better."[101]

Meanwhile, the hobos sung songs of a world turned upside down. The original lyrics to the song "Big Rock Candy Mountain" by Harry McClintock, also known as "Haywire Mac," depicted a world where the collectivist values of the periphery triumphed over the proprietary values of the center: authority was hamstrung, economy absent, and leisure abundant. There was justice and transparency: "a land that's fair and bright." Evading the law was easy because "all the cops have wooden legs," and "the jails are made of tin, and you can walk right out again as soon as you are in." The commons provided enough for all: "the hens lay soft-boiled eggs, the farmer's trees are full of fruit and the barns are full of hay," and "the little streams of alcohol come a-trickling down the rocks." Work was absent and leisure abundant: "There ain't no short-handled shovels, no axes, saws or picks, I'm a-goin' to stay where you sleep all day."[102]

McClintock's song harkens back to a long tradition of distant lands where hierarchical authority is nonexistent and labor is foreign. The "Golden Age" associated with the rule of Kronos in Hesiod's "Works and Days" was characterized by such conditions. Hesiod observed that "the Gods desire to keep the stuff of life hidden from us. If they did not, you could work for a day and earn a year's supplies."[103] Pherecrates extrapolated in the fourth century BC with verse, describing "rivers full of porridge . . . sliced fish all cooked . . . filled cups. . . . And if anyone ate or drank of these things, twice as much as there was first spurted up again."[104] The idea of life without work persisted from the ancient to medieval epochs. In thirteenth-century Europe, songs and poems of the Land of Cockaigne imagined a place where "every man takes what he will, as of right to eat his fill. All is common to young and old, to stout and strong, to meek and bold."[105]

Technocratic monoculture asserts the apparatus of security as the manifestation of spatial stasis and hierarchical control. The anonymous authors of *The Coming Insurrection* observed that "freedom is no longer a name scrawled on walls, for today it is always followed, as if by its shadow, with the word 'security.'"[106] This applies to people as well as institutions. The laborer not bound by work contract and mortgage is problematic; his or her freedom is absent of any economic obligation to the institutions of capitalism. David Harvey observes that the mortgage interest income tax deduction was designed to get workers at the turn of the twentieth century to "buy in" to the system, and that their financial obligations would render their unions less militant, as workers would be less inclined to strike for fear of losing their homes.[107] Mobility implies freedom. From native populations, to hobos, gypsies, bikers, and "illegal aliens," the suspicion that these groups are operating outside of the bounds of capitalist modernity has historically made them targets of social ostracization, political recrimination, and economic penalty.

At the same time, capitalist modernity commodifies the experience of nomadism through recreation. Unlike its original form, nomadism in modern society is accessible only through investment. To recreate participation in a nomadic lifestyle, one must buy in. Motorcycles, RVs, hiking equipment, and the necessary official permits and licenses to participate in such activity ensures that modern recreations of nomadism reinforce the expectations of capitalism, not challenge it. They are temporary escapes, always with a return destination the same as the initial point of departure, temporary loops that serve as pressure release valves from the stresses of modernity for those who can afford to participate. In *The Society of the Spectacle*, Guy Debord suggested that:

> In the expanding economy of 'services' and leisure activities, the payment for these blocks of time is equally unified: 'everything's included,' whether it is a matter of spectacular living environments, touristic pseudotravel, subscriptions to cultural consumption or even the sale of sociability itself. . . . Spectacular commodities of this type . . . would obviously never sell were it not for the increasing impoverishment of the realities they parody . . .[108]

Their presence stands in defiance of the technocratic monoculture's attempt to impose a standardized, ontological order over the *longue durée* of capitalist globalization. The result is ontological stasis, not teleological change. The panoptics of spatial command and control, as the authors of *The Coming Insurrection* suggest, become more pervasive and complete within the enabling construct of modernity: "As an attempted solution, the pressure to ensure that nothing happens, together with police surveillance of the territory, will only intensify. . . . The territory will be partitioned into ever more restricted zones. Highways built around the borders of 'problem neighborhoods' already form invisible walls closing off those areas from the middle class subdivisions."[109] *The pressure to ensure that nothing happens.* Ontological stasis, enforced through spatial codes and barriers, protects the economic beneficiaries of technocratic monoculture by halting the teleological intention to undo it.

ASSIMILATION

With the triangulation of space and the imposition of security comes the project of assimilation. Assimilation is vital to the expansive nature of technocratic monoculture. The development of an American infrastructure went hand in hand with the development of an American identity. Central to the construction of that identity was the process of assimilation. As large numbers of immigrants came to the United States for work in the expanding industrial economy, they were encouraged by the state and corporations to shed their ethnicity in exchange for cultural acceptance. Certain groups, like the Germans and Irish for example, were able to do this based on their physical appearance, which was similar enough to the Anglo-Saxon identity that had come to define American racial standards for acceptance, but other groups were less successful. The Chinese, who contributed greatly to the construction of the western infrastructure, were marginalized economically, politically, and culturally due in large part to their physical appearance, which differentiated them from the constructed standard of American identity and rendered them second class citizens in the process.

The idea of cultural assimilation as an aspect of technocratic monoculture is best represented by the turn of the twentieth century concept of "The Melting Pot." Israel Zangwill popularized the concept in his 1908 play of the same name. In the play, the main character is Russian-born David Quixano, who is the sole surviving member of his family that was killed in the 1903 Kishinev Pogrom. David emigrates to America to write a symphony called "The Crucible," which he intends to be a musical representation of the possibilities of American identity. David explains his artistic intent to his girlfriend Vera at the end of the play:

> It is the fires of God round His Crucible. There she lies, the great Melting Pot—listen! Can't you hear the roaring and the bubbling? There gapes her mouth—the harbour where a thousand mammoth feeders come from the ends of the world to pour in their human freight. Ah, what a stirring and a seething! Celt and Latin, Slav and Teuton, Greek and Syrian,—black and yellow—Jew and Gentile.
>
> Yes, East and West, and North and South, the palm and the pine, the pole and the equator, the crescent and the cross—how the great Alchemist melts and fuses them with his purging flame! Here shall they all unite to build the Republic of Man and the Kingdom of God. Ah, Vera, what is the glory of Rome and Jerusalem where all nations and races come to worship and look back, compared with the glory of America, where all races and nations come to labour and look forward! Peace, peace, to all ye unborn millions, fated to fill this giant continent—the God of our *children* give you Peace.[110]

Zangwill's play was both popular and successful at the time, prompting President Theodore Roosevelt to exclaim at the conclusion of its opening night in Washington, D.C.: "That's a great play, Mr. Zangwill!" Critics, though, were less impressed. *The London Times* critic A. B. Walkley dismissed the play as "romantic claptrap" and dismissed Roosevelt's enthusiasm for the performance as "stupendous naivete."[111] Nevertheless, its popularity among its target audience: working class immigrants looking for validation in the idea that they could become part of something progressive and transcendent, was enough to make the play a commercial success. The conflation of economic gain and lofty national ideals contributed to the popular belief in the concept of "The Melting Pot."

Increasingly though, the identity of this national crucible was being forged not by the political and ideological forces of nationalism, but by the economic forces of corporate capitalism. The corporate appropriation of American identity as it relates to the immediate concept of "The Melting Pot," and the overarching concept of technocratic monoculture, is evident in the institution of the Ford English School, which sought to use the idea to create docile, passive workers and consumers rather than engaged, proactive citizens. The Ford English School was established in 1914 to cloak corporate interests in the construction of a patriotic American identity. The English School was operated by the Ford Sociological Department, a wing of the corporation charged with the task to:

> explain opportunity, teach American ways and customs, English language, duties and citizenship . . . counsel and help unsophisticated employees to obtain and maintain comfortable, congenial and sanitary living conditions, and . . . exercise the necessary vigilance to prevent, as far as possible, human frailty from falling into habits or practices detrimental to substantial progress in life.[112]

The Ford Sociological Department, of which the English School was an auxiliary, was reflective of a larger social agenda put forth the Detroit Americanization Committee. Founded in 1915 by the Detroit Chamber of Commerce, it included six representatives from Detroit's largest corporations, including John R. Lee, director of the Ford Sociological Program.[113] Here the interests of capital shaped the idea of American identity: English Language courses at the Ford School used vocabulary lessons to shape workers' identities as consumers of American goods and services, and as producers deferential to the hierarchical corporate structure. Lessons like "Going To The Bank," "Beginning The Day's Work," and "Finishing The Day's Work" had less to do with forging the identity of civically minded citizens than with shaping workers identities as deferential consumers. Its design was intended to mimic the assembly line production of automobiles. In the words of Educational Department head Samuel S. Marquis, "this is the human mind we seek to turn out, and as we adapt the machinery in the shop to turning out the kind of automobile we have in mind, so we have constructed our educational system with a view to producing the human product in mind."[114]

Graduation from the Ford English School was perhaps the most tangible example of Marquis's statement regarding the ultimate goal of the project. One of the school's subsequent directors, Clinton C. Dewitt, described the ritual in which graduating students descended into a giant "melting pot" in the dress of their nationality, and emerged uniformly dressed as Americans, waving flags, symbolizing their transformation:

> a pageant in the form of a melting pot, where all the men descend from a boat scene representing the vessel on which they came over, down a gangway, into a pot 15 feet in diameter and 7 ½ feet high, which represents the Ford English School. . . . Into the pot 52 nationalities with their foreign clothes and baggage go and out of the pot after vigorous stirring by one of the teachers comes one nationality, viz, American.

The philosopher Horace Kallen openly challenged the idea of assimilation in his 1915 work *Democracy versus The Melting Pot*, arguing for a form of cultural pluralism in American identity which defied corporate expectations that national identity be a form of economic control. Kallen argued that "until the disparity between our economic resources and our population becomes equalized, so that the country shall attain an approximate economic equilibrium," any veneer of "Americanization" was superficial. Assimilation could not obfuscate class difference, and was a social policy essentially designed to placate immigrant workers who were being "submerged beneath a conquest so complete that the very name of us means something not ourselves."[115]

The idea of a Melting Pot, which promoted a political and cultural unity, was widely abandoned in the 1990s when the economic unity of the underclasses became a driving force of teleological change in the wake of Reagan's economic policies. Multiculturalism, which highlights and celebrates the ethnic differences of Americans, stands in sharp contrast to the downplaying and minimizing of difference for most of America's history. With this, the unity of purpose espoused by corporately defined nationalism was exchanged in favor of difference, allowing for a plurality of market demographics that offered corporate capitalism a range of superficial options to provide consumers with which to express their "individuality" through participation in the market economy. Multiculturalism that occurs in a capitalist framing is ultimately monoculture. Again, with The Invisible Committee: *"the West has sacrificed itself as a particular civilization to impose itself as a universal culture.* The operation can be summarized like this: an entity in its death throws [sic] sacrifices itself as a content to survive as a form."[116]

TRIANGULATION

The conditions of technocratic monoculture are ever more pervasive and expansive. At the same time, the detrimental economic, political, and social effects of technocratic monoculture are increasingly apparent. Technocracy, rooted in surveillance and market economics, commodifies everything, including the very democratic politics that would inhibit its further growth and expansion. The result is an economic system based on exploitation. Monoculture, in the agricultural and social sense of the word, is ecologically and sociologically destructive. Applied spatially, technocratic monoculture imposes the logic of security to usurp individual and collective freedoms, resulting in the comprehensive forms of assimilation necessary for its continued expansion. The result is a condition that The Invisible Committee observes at the very opening of *The Coming Insurrection*: "From whatever angle you approach it, the present offers no way out."[117]

A multidisciplinary team of academics produced a paper in 2012, partially funded by the NASA Goddard Institute for Space Studies, titled *A Minimal Model for Human and Nature Interaction*, in which they observe that "over-exploitation of natural resources and strong economic stratification—can result in complete collapse [of modern civilization]."[118] Despite the ominous warning signs, technocratic monoculture continues to expand its scope and reach. Such collective behavior is a phenomenon psychologist Barry M. Staw identified in individuals in his 1976 paper *Knee Deep in the Big Muddy: A Study of Escalating Commitment to a Chosen Course of Action*. Staw observed that "it is commonly expected that individuals will reverse decisions or change behaviors which result in negative consequences. Yet, within investment decision contexts, negative consequences may actually cause decision makers to increase the commitment of resources and undergo the risk of further negative consequences."[119] It would seem that what applies to the individual agents in Staw's study also holds true within the comprehensive timespace of modernity.

NOTES

1. The Invisible Committee, *The Coming Insurrection* (Los Angeles: Semiotext(e), 2009), 16.
2. Fredy Perlman, *Against His-Story, Against Leviathan!* (Detroit: Black & Red, 1983), 42.
3. James Clifford, "Traveling Cultures" in *Cultural Studies*, eds. Lawrence Grossberg, Cary Nelson, and Paula Treicher (New York: Routledge, 1992), 110.
4. Alan R. Drengson, "Shifting Paradigms: From Technocrat to Planetary Persons," in *The Deep Ecology Movement: An Introductory Anthology*, ed. Alan R. Drengson and Yuichi Inoue (Berkeley: North Atlantic Books, 1995), 84.
5. Angela Davis, "Masked Racism: Reflections on the Prison Industrial Complex," *Color Lines*, Sept 10, 1998, http://colorlines.com/archives/1998/09/masked_racism_reflections_on_the_prison_industrial_complex.html
6. John Howard "An Account of the Principal Lazerettos of Europe," *The Works of John Howard, Esq. Vol. II* (London: Johnson, Dilly and Cadell, 1791), 226–7.
7. John Howard, *The State of the Prisons in England and Wales with Preliminary Observations, and an Account of Some Foreign Prisons* (Warrington: William Eyres, 1777), 72.
8. Jeremy Bentham, *The Panopticon Writings*, ed. Miran Bozovic (London: Verso, 1995), 31.
9. Ibid. 43.
10. Ibid. 56.
11. Michel Foucault, *Discipline and Punish: The Birth of the Prison* (New York: Penguin, 1977), 172.
12. Edmund F. Byrne, *Public Power, Private Interests and Where to We Fit In?* (New York: 1st Book Library, 1998), 4.
13. Jeremy Bentham, *The Panopticon Writings*, ed. Miran Bozovic (London: Verso, 1995), 31, 50.
14. Ibid. 51.
15. Andrew Bard Schmookler, *The Illusion of Choice: How the Market Economy Shapes Our Destiny* (Albany: State University of New York Press, 1993), 25.
16. David Levine, "Recombinant Family Formation Strategies." *Journal of Historical Sociology*, Vol. 2, No. 2 (1989): 106.
17. Benjamin Barber, *Jihad vs. McWorld: Terrorism's Challenge to Democracy* (London: Corgi, 2003), xii.

18. Karl Marx, *Capital: A Critical Analysis of Capitalist Production* (London: Swann & Sonnenschein & Co., 1902), 1.
19. Jennifer Lee, "EPA Said to Avoid Studies Conflicting White House," *New York Times*, July 14, 2003.
20. "Pinsent-Masons Launches 2010–11 Water Industry 'Bible,'" accessed June 2, 2014, www.pinsentmasons.com/en/media/press-releases/2010/pinsent-masons-launches-2010–11-water-industry-bible/
21. Steve Maxwell, *A Concise Review of Challenges and Opportunities in the World Water Market*, (2012): 4, 28. www.summitglobal.com/documents/Maxwell2012WaterMarketReview-a030912.pdf
22. Will Henley, "What Value Should We Place on Water in Developing Countries," *The Guardian*, September 17, 2013. www.theguardian.com/sustainable-business/value-water-developing-countries-talkpoint
23. David Harvey, *A Brief History of Neoliberalism* (Oxford: Oxford University Press, 2005), 165.
24. Steve Charnovitz "Addressing Environmental and Labor Issues in the World Trade Organization," *Trade and Global Markets: World Trade Organization*. Progressive Policy Institute. November 1, 1999, www.dlc.org/ndol_cid308.html?kaid=108&subid=128&contentid=649
25. Janet Semple, *Bentham's Prison: A Study of the Panopticon Penitentiary* (Oxford: Oxford University Press, 1993), 152.
26. Michel Foucault, *Discipline and Punish: The Birth of the Prison* (New York: Penguin, 1977), 149.
27. The Invisible Committee, *The Coming Insurrection* (Los Angeles: Semiotext(e), 2009), 16.
28. The Royal Academy. "The Eighty-Seventh Exhibition, 1855," *The Art-Journal*. (June 1, 1855): 21.
29. William Henry Smyth, "Technocracy—National Industrial Management," in *Technocracy*, (Berkeley: W.H. Smyth, 1920), 13.
30. Ibid. 34
31. Thorstein Veblen, *The Engineers and the Price System* (New York: B.W. Heubsch, 1921), 33–34.
32. Ibid. 158.
33. Howard Scott, *Technocracy: Science vs. Chaos* (Akron: Technocracy Inc., 1933), 10–21.
34. Benjamin DeCasseres, "The Hit and Run Thinker: Technocracy Analyzed and Frisked," *Life Magazine*, Feb. 1933, 100.
35. John Langdon-Davies, "Ramie: King Cotton's Rival." *The Forum* Vol. 90 (November 1933): 289.
36. Howard P. Segal, *Technological Utopianism in American Culture* (Syracuse: Syracuse University Press, 2005), 123.
37. Howard Scott, *Technocracy: Science vs. Chaos* (Akron: Technocracy Inc., 1933), 15.
38. C. L. Sulzberger, "Europe Asks Questions on U.S. Foreign Policy," *The New York Times*, May 11, 1952, E3.
39. W. Granger Blair, "Newcastle Rises From the Doldroms: Once Backward English City is Being Transformed," *The New York Times*, February 27, 1966.
40. Lloyd Garrison, "Nigeria Regains Business Capital," *The New York Times*, May 15, 1966.
41. Frank Fischer, *Technocracy and the Politics of Expertise* (New York: Sage, 1990), 145.
42. Charles M. Fox, *Psychopathology of Hysteria* (Boston: The Gorham Press, 1913), 5, 33.
43. Richard L. Tames, ed., *Documents of the Industrial Revolution 1823* (London: Hutchinson, 1971), Christian P.W. Bleuth to K.F. Schinkel, 69.

44. Lucy Larcom, *A New England Girlhood, Outlined from Memory* (New York: Houghton and Mifflin, 1889), 183.
45. Gerald Walsh, *Industrialization and Society: Selected Sources* (London: McClelland and Steward, 1969), 48.
46. Robert Gottlieb, *Environmentalism Unbound: Exploring New Pathways For Change* (Boston: MIT Press, 2002), 21.
47. Charles Wing, *The Evils of the Factory System, Demonstrated by Parliamentary Evidence* (London: Frank Cass & Co., 1967), 191.
48. Friedrich Engels, *The Condition of the Working-Class in England in 1844* (London: Swan Sonnenschein & Co., 1892), 48–53.
49. Charles S. Myers, "A Contribution to the Study of Shell Shock," *The Lancet*, February 13, 1915, 316.
50. Craig Brod, "Managing Technostress: Optimizing the Use of Computer Technology." *Personnel Journal*, Vol. 61 (October 1982): 753–757.
51. Chaim F. Shatan, "Post-Vietnam Syndrome," *New York Times*, May 6, 1972.
52. Hans Binneveld, *From Shellshock to Combat Stress: A Comparative History of Military Psychology* (Amsterdam: Amsterdam University Press, 1997), 191.
53. "DSM-5 Diagnostic Criteria for PTSD Released," United States Department of Veterans Affairs, National Center for PTSD, December 5, 2012. www.ptsd.va.gov/professional/pages/diagnostic_criteria_dsm-5.asp
54. "Post Traumatic Stress Disorder," Nebraska Department of Veterans Affairs, 2007. www.ptsd.ne.gov/what-is-ptsd.html
55. Perry Anderson, *Lineages of the Absolutist State* (London: Verso, 1974), 283–285.
56. Ibid. 428–429
57. Richard B. Sheridan, *Sugar and Slavery: An Economic History of the British West Indies, 1623–1775* (Jamaica: Canoe Press, 1974), 124.
58. Ibid. 140.
59. Ibid. 254–255.
60. Gary Snyder, *The Real Work: Interviews and Talks, 1964–1979* (New York: New Directions Books, 1980), 160.
61. David Harvey, *The Condition of Postmodernity: An Enquiry into the Origins of Cultural Change* (Cambridge: Blackwell, 1989), 240.
62. Anthony Giddens, *The Consequences of Modernity* (Cambridge: Polity Press, 1990), 14–20.
63. Jonathan Friedman, "Being in the World: Globalization and Localization," in *Global Culture: Nationalism, Globalization and Modernity*, ed. Mike Featherstone (London: Sage Publications, 1990), 311.
64. Barbara Bradby, "The Destruction of the Natural Economy." *Economy and Society*, Vol. 4, No. 2 (1975): 127.
65. Naomi Klein, *The Shock Doctrine* (New York: Penguin, 2008), 17–18.
66. Karl August Wittfogel, *Oriental Despotism: A Comparative Study of Total Power* (New Haven and London: Yale University Press, 1957), 1.
67. Christopher Francese, *Ancient Rome in So Many Words* (New York: Hippocrene, 2007), 79.
68. Pliny, *The Natural History of Pliny*, trans. John Bostock and H. T. Riley (London: Henry G. Bohn, 1856), Vol. 4, 14, Book XVIII, 7.
69. Ibid. Vol. 4, 15, Book XVIII, 7.
70. Rodney Hilton, *Class Conflict and the Crisis of Feudalism: Essays in Medieval Social History* (London: Hambledon Press, 1985), 140–141.
71. Donald Worster, *Rivers of Empire: Water, Aridity and the Growth of the American West* (New York and Oxford: Oxford University Press, 1985), 51.
72. Ibid. 52.
73. "Influence and Lobbying: Agribusiness," Center for Responsive Politics, accessed June 2, 2014. www.opensecrets.org/industries/indus.php

74. James Howard Kunstler, *The Geography of Nowhere: The Rise and Decline of America's Man Made Landscape* (New York: Free Press, 1994), 93–94.
75. Plato, *Timaeus*, trans. Benjamin Jowett (Rockville: Serenity, 2009), 115.
76. Eric G. Forbes, *Greenwich Observatory: One of Three Volumes by Different Authors Telling the Story of Britain's Oldest Scientific Institution* (London: Taylor and Francis, 1975), Vol. 1, 22.
77. Reginald Horsman, "The Dimensions of an 'Empire of Liberty': Expansion and Republicanism, 1775–1825." *Journal of the Early Republic*, Vol. 9, No. 1 (Spring 1989): 24–25.
78. Gregory Nobles, "Straight Lines and Stability: Mapping the Political Order of the Anglo-American Frontier." *The Journal of American History*, Vol. 80, No. 1 (June 1993): 22–26.
79. Thomas Jefferson, "Memoir of Meriweather Lewis," in *The History of the Lewis and Clark Expedition*, ed. Elliott Coues (New York: Francis P. Harper, 1893), Vol. 1, xxvi.
80. Frederick Jackson Turner, *The Frontier in American History* (New York: Frederick Holt & Co., 1920), 1.
81. George Price, *A Treatise on Fire and Thief Proof Depositories and Locks and Keys* (London: Simpkin, Marshall and Co., 1856), 8–9.
82. Ibid. 25
83. Ibid. 57–58.
84. Granville Sharp, *The Gilbart Prize Essay on the Adaption of Recent Discoveries and Inventions in Science and Art to the Purposes of Practical Banking* (London: Groombridge and Sons, 1854), 291.
85. Ibid. 291
86. "Groundbreaking Study Finds U.S. Security Industry to be $350 Billion Market; Integrators Vital to End Users' Decision-Making," *SDM Magazine*, August 20, 2013, www.sdmmag.com/articles/89562-groundbreaking-study-finds-us-security-industry-to-be-350-billion-market-integrators-vital-to-end-users-decision-making
87. Ed Harris and Frank Sammartino, "Trends in the Distribution of Household Income, 1979–2009," Congressional Budget Office, August 6, 2012, www.cbo.gov/sites/default/files/cbofiles/attachments/Trends_in_household_income_forposting.pdf
88. Edward James Blakely and Mary Gail Snyder, *Fortress America: Gated Communities in the United States* (Washington: The Brookings Institution, 1997), 3–5.
89. Ibid. 8.
90. Ibid. 8.
91. Daniel D'Oca, "The Arsenal of Inclusion and Exclusion." *MAS Context*, No. 17, (Spring 2013): 70.
92. J. R. Wordie, "Chronology of English Enclosure, 1500–1914." *The Economic History Review*, Vol. 36, No. 4 (November 1983): 483.
93. Sir Thomas More, *Utopia* (London: Richard Chiswell, 1684), 11–12
94. Peter Linebaugh and Marcus Rediker, *The Many Headed Hydra: The Hidden History of the Revolutionary Atlantic* (Boston: Beacon Press, 1999), 18.
95. Karl Marx, *Capital: A Critical Analysis of Capitalist Production* (London: Swan & Sonnenschein & Co., 1902), 761.
96. Rev. Frank Laubach, *Why There Are Vagrants: A Study* (New York: Columbia, 1916), 42.
97. Ibid. 42.
98. Kenneth L. Kusmer, *Down and Out, the Homeless in American History* (Oxford: Oxford University Press, 2002), 74–75.
99. John C. Schneider, *Detroit and the Problem of Order* (Lincoln: University of Nebraska Press, 1980), 126–127.

100. Nels Andersen, *The Milk and Honey Route: A Handbook for Hobos* (New York: Vanguard Press, 1931), 21.
101. Kenneth L. Kusmer, *Down and Out, the Homeless in American History* (Oxford: Oxford University Press, 2002), 41.
102. Harry McClintock, "The Big Rock Candy Mountains," Victor #21704, 1928, http://archive.org/details/TheBigRockCandyMountains
103. Hesiod, "Works and Days," in *Hesiod and Theogonis*, trans. Dorothea Wender (New York: Penguin Books, 1973), 60.
104. Athaneaus, *Deipnosophistae* (New York: Loeb Classical Library, 1929), Book VI, 20.
105. J. C. Davis, *Utopia and the Ideal Society: A Study of English Utopian Writing 1516–1700* (Cambridge: Cambridge University Press, 1981), 21.
106. The Invisible Committee, *The Coming Insurrection* (Los Angeles: Semiotext(e), 2009), 56.
107. Vince Emanuele, "Interview with David Harvey: Rebel Cities and Urban Resistance Part II," *Z Magazine*, January 7, 2013, www.zcommunications.org/contents/190562
108. Guy Debord, *The Society of the Spectacle* (Canberra: Treason Press, 2002), 43.
109. The Invisible Committee, *The Coming Insurrection* (Los Angeles: Semiotext(e), 2009), 16.
110. Israel Zangwill, *The Melting Pot* (London: William Heineman, 1914), 184–185.
111. Maurice Wohlgelernter, *Israel Zangwill: A Study* (New York: Columbia University Press, 1964), 176–177.
112. Stephen Meyer III, *The Five Dollar Day: Labor Management and Social Control in the Ford Motor Company, 1908–21* (Albany: State University of New York Press, 1981), 127.
113. Clarence Hooker, *Life in the Shadows of the Crystal Palace, 1910–27: Ford Workers in the Model T Era* (Bowling Green: Bowling Green State University Popular Press, 1997), 115.
114. Stephen Meyer III, *The Five Dollar Day: Labor Management and Social Control in the Ford Motor Company, 1908–21* (Albany: State University of New York Press, 1981), 156–9.
115. Horace Kallen, "Democracy versus the Melting Pot: A Study of American Nationality," *The Nation*, Feb. 25, 1915.
116. The Invisible Committee, *The Coming Insurrection* (Los Angeles: Semiotext(e), 2009), 60.
117. Ibid. 13.
118. Safa Motesharrei, Jorge Rivas, and Eugenia Kalnay, "A Minimal Model for Human and Nature Interaction," November 13, 2012, www.ara.cat/societat/handy-paper-for-submission-2_ARAFIL20140317_0003.pdf
119. Barry M. Staw, "Knee Deep in the Big Muddy: A Study of Escalating Commitment to a Chosen Course of Action." *Organizational Behavior and Human Performance*, Vol. 16, Issue 1 (1976): 27.

4 Of Spectacles and Monuments

INTRODUCTION

Theogony begins at "the altar of the mighty Zeus" on Mount Helicon, with Hesiod explaining that the Muses "breathed a sacred voice into my mouth with which to celebrate the things to come and the things that were before," and then "ordered me to sing the race of blessed ones who live forever."[1] The altar is a monument at which the commemoration of the mythological spectacle that was the Titanomachy takes place in the story. These elements, spectacle, monument, and commemoration, are essential to the ontological control of space that maintains temporal stasis and prevents teleological change from occurring. Controlling the memory of past events through the construction and scripting of monumental architecture serves the function of shaping public memory in a way that serves to reinforce the hierarchical power structures of modernity. By separating the past from the present through monumental commemoration, and reinforcing the temporal disconnect through ritualized reverence that inherently produces memory of past events in ways that serve to reinforce the interest of technocratic monoculture, consolidated hierarchical political and economic power constructs itself as the inheritor and protector of revolutionary ideology, rather than its primary target.

This chapter explores these arguments by first examining the conceptual framing of events as symbols of historical change, then by considering how such events, in a teleological context, represent a form of chaos to the ascendant hierarchy of modernity which must be ordered, regulated, and controlled. In the aftermath of historical events that advance teleological progress, the technocratic monoculture assumes control of the narrative by fixing the temporal event in space through the construction of monumental architecture, which serves to appropriate the memory of the event and shape its commemoration so that it serves to reinforce ontological stasis rather than challenge it. To demonstrate this point, this chapter considers examples of monuments that commemorate historical events that were intended to challenge modern forms of hierarchy and hegemony, and considers how those events have been abstracted and reshaped to serve the technocratic monoculture they were intended to subvert. Lastly, the chapter explores the imperialist appropriation of space by examining the U.S. military's counter-insurgency policy of "Clear Hold Build," and explores how such policies have been used over the course of history to reshape and produce space to serve the political, economic, and social interests of modern technocratic monoculture.

EVENTS AS SYMBOLS OF HISTORICAL CHANGE

If we are to accept the premise that time is measured by events that symbolize historical change, then we need to consider the context in which these changes occur and how historians have interpreted them in the modern era. The Belgian-born U.S. literary critic Paul DeMan wrote that "modernity exists in the form of a desire to wipe out whatever came earlier, in the hope of reaching at least a point that could be called a true present, a point of origin that marks a new departure."[2] Similarly, Modris Eksteins wrote in his account of World War I that modernism is "the culture of the sensational event."[3] The idea that sensational events are pivotal in the historical narrative seems matter of fact. Think of the introductions to so many historical documentaries. They are often an amalgamation of historical events caught on film: the Atomic Bomb footage from Los Alamos, Martin Luther King, Jr. speaking on the National Mall in Washington, D.C., the Moon Landing, Nixon waving from Marine One, German citizens tearing down the Berlin Wall, and so on. The essential nature of change in our understanding of historical time is self-evident. We articulate change in elemental terms: groundbreaking events (earth), flashpoints in history (fire), winds of change (air), and watershed moments (water), and measure history by a series of temporal monuments over a timescape to reckon where society is by observing where it has been. As Alvin Toffler explained, "without time, change has no meaning. And without change, time would stop. Time can be conceived as the intervals during which events occur."[4]

The Situationists understood the power of the spectacle. Formed in 1957, the European group of intellectuals and social revolutionaries drew from artistic and academic theory to comprehensively critique the condition of modernity. Guy Debord wrote in *The Society of the Spectacle* that "in societies where modern conditions of production prevail, life is presented as an immense accumulation of *spectacles*." Spectacles, in Debord's opinion, were "a means of unification," which were ultimately "a visual reflection of the economic order." The result of the commemoration of spectacle was separation and passivity, which ultimately "separates what is *possible* from what is *permitted*." Modern monuments commemorate the spectacle of revolutionary actions, but while they are celebrated in the past tense, revolutions are not permitted in the present. The result, according to Debord, was that "the spectacle is the bad dream of a modern society in chains, and ultimately expresses nothing more than its desire to sleep. The spectacle is the guardian of this sleep."⁵ The quote is evocative of Francisco Goya's 1799 etching *The Sleep of Reason Produces Monsters*, in which the artist, slumped over at his drawing table, is flanked by darkness and nocturnal animals who threaten to overrun the frame (See Figure 4.1). Goya's print was part of a series which critiqued Spanish culture at a time when Enlightenment thought was offering a fundamental challenge to the ordering of political and socioeconomic hierarchy throughout the modern world. Both Goya and Debord encouraged a critical awareness that the hierarchical political order actively discourages and seeks to undermine. Debord considered the dreams of the sleeping, which lay dormant in the waking consciousness of technocratic monoculture, "the temporal realization of authentic communism," teleological liberation, or a return to Hesiod's "Golden Age" under the reign of Kronos. He concluded that "the world already dreams of such a time. In order to actually live it, it only needs to become fully conscious of it."⁶

The Situationists asserted that a society subject to the spectacle is separated from its own agency. What is possible is not permitted. Revolutions no longer occur; they are studied and commemorated. The fighting *has been done* (emphasis on the passive voice). By whom, though? By *patriots*: people who, by definition, love their country. The paradox is obfuscated and enforced simultaneously. The historicization of the American Revolution serves as an expedient model of this phenomenon. In the American Periodicals database, which makes accessible the full text of over 1,100 American magazines, journals, and newspapers from 1741–1900, the term "patriot" was used sixty-two times during the Revolutionary Era (1763–1783) and 63,773 times from 1800–1883. While surely some of this growth can be accounted for by the expansion of the print industry and what Jurgen Habermas calls "the public sphere," much of that growth is also a result of the fact that we write and shape our history from a safe distance. This is not to say that the concept of patriotism was anachronistic to late eighteenth-century America; rather, the idea of patriotism during the period was both institutionalized and romanticized afterwards. From a safe temporal distance, the narrative of patriotism was shaped and framed by the architects of a comprehensive political, economic and social system that disenfranchised entire cross sections of the American population. While the political design suppressed and challenged inclusive democratic participation in government, the economic system subjugated the population to the machines of the Industrial Revolution. and socially ostracized them by problematizing and outright criminalizing their culture and identities. In the meantime, revolution is relegated to the past tense. It is something that has happened, and something to be grateful for, but not something to replicate or otherwise assert in the present tense. Revolutionary history is romanticized, but current revolutionary ideas are criminalized as obscenity.

The ultimate spectacle in the linear construction of modern history has yet to occur. The apocalypse, as written in the Book of Revelation and reinforced

Figure 4.1 The Sleep of Reason Produces Monsters (*El sueño de la razon produce monstrous*).

by Francisco Goya, 1797–1799. Museo de Prado via Wikimedia Commons. http://upload.wikimedia.org/wikipedia/commons/f/f5/Museo_del_Prado_-_Goya_-_Caprichos_-_No._43_-_El_sue%C3%B1o_de_la_razon_produce_monstruos.jpg.

in cultural narrative since its conception, is foretold in Christianity as the last and greatest spectacle of all. Its presence on the temporal horizon serves as an ideological brake on the teleological progress articulated by Hegel and Marx. In 1948, Evgeni Lampert wrote in *The Apocalypse of History* that, within an eschatological framing, "our History becomes not merely a series of happenings but the disclosure and consummation of divine and human destiny, that is, apocalypse."[7] To prevent the end times from coming, hierarchical political and economic institutions enforce stasis. Herein lays the tension between temporal change and spatial stasis. Again, with Lampert, on the connection of past and present through spectacle: "Admittedly, as bare 'fact' or 'datum' the ways of men pertain to the irresistible flux of Time; but as event they are charged with transcendent power and actuality and are intensely valid and present here and now, however much they may be removed from us in Time and Space. They are the situation, and we cannot abstract ourselves from it." He concluded that "the idea of evolution was, undoubtedly, a fully justified reaction against such a glorification of static immutability. . . ."[8] Any change is a progression toward apocalypse in the ontological world view; temporal stasis puts it off indefinitely. As such, eschatology serves as the panic button for the preservation of hierarchical order, which can be threatened at any attempt to challenge ontological stasis. "Static societies," as Guy Debord explained, "are societies that have reduced their historical movement to a minimum and that have managed to maintain their internal conflicts and their conflicts with the natural and human environment in a constant equilibrium. In these societies a definitive organizational structure has eliminated any possibility of change."[9]

Spectacular events, meant to advance teleology toward a Hegelian concept of progress then, are co-opted in the modern epoch by the state and the market, and their meanings are reengineered to reinforce the very structures they were meant to challenge. Hegel argued that human history exposed a type of progress that moved toward "the realization of spirit."[10] The appropriation of spectacular events, through spatial and ideological commemoration, thwarts this temporal progress. In the process, the past and the present are severed from each other; present-day onlookers fail to recognize their own reflection in the statues commemorating the agents of teleological change, because their images have been rendered unrecognizable. Here again, Debord elucidated the situation: "this society produces a new internal distance in the form of spectacular separation."[11] Controlling the spatial and ideological commemoration of spectacular events is critical to the maintenance of ontological temporal stasis.

Modernity's attendant systems, technology, the dual forces of nationalism and urbanization, and market economics, work to undermine any attempt at breaking free of this stasis, holding humanity subject, captive, exposed, and exploited at its own expense. Technology allows the constructed memory to proliferate unhindered, nationalism sanctions the constructed memory in a way that reinforces rather than challenges systematic control, while urbanization provides a spatial context for monumental significance and market economics obscure the meaning of holidays that commemorate spectacular events with the promise of increased access to commodified goods and services. Democratic ideology is marginal in the space of commemoration. Its theory remains vital to the meaning and significance of commemorated spectacular events in an abstracted and distanced way, but its practice is often met with police barricades and court dates. Again, with Debord: "The bourgeoisie has thus made irreversible historical time known and has imposed it on society, but has prevented society from *using* it." In the process, the constructed meaning of the spectacle, which reinforces state and capitalist control of timespace and ideology, is amplified and reified "with the preservation of a new immobility *within history*."[12]

CHAOS AND ORDER

A fundamental tension in the contest between teleological progress and spatially enforced ontological stasis is trenched in ideas of chaos and order. Hierarchical order is imposed on space to eliminate variables that would threaten the status quo; those variables are identified and constructed as chaotic and threatening within the confines of the technocratic monoculture. Those forces that seek to undo the hierarchically imposed order by challenging the spatial hegemony of the monoculture as a way to move teleological progress forward threaten the security of the state. Here, Chaos, the prime mover in Hesiod's telling of the Titanomachy, from which Earth itself comes (geographical space), as well as Tartarus (the Underworld), and Day and Night (markers of temporal progress), is threatened by the spatial ordering of Olympian power. From Chaos comes both time and space. Time harbors freedom from below, spatial ordering seeks to control the space in which it would manifest to promote security: Chaos against order, freedom against security.

The tension between these metaphysical forces manifests throughout myth and literature over the course of human history. In the last lectures he gave before his death in 2004, Jacques Derrida began to articulate the literary war the two forces have waged in a series of lectures called *The Beast and the Sovereign*. One of his fundamental points was that both exist beyond the range of the law. The sovereign is above it, while the beast remains outside of it.[13] While the sovereign's position above the law provides him or her with the means to impose arbitrary will for purposes of maintaining hierarchically imposed order and control, the beast's position outside of the law enables him or her to challenge the authority of the sovereign and withdraw from striking distance accordingly.

While Derrida's analysis largely focuses on fictional literary accounts of this phenomenon, it is an argument that applies to this analysis. Throughout the course of history, sovereign powers have engaged in extralegal activities to maintain hierarchically imposed order when it was threatened by outlaws existing on the peripheries of their dominion. As a result, the extension of spatial control, through the development of mapping and surveillance technologies, became an integral part of the burgeoning technocratic monoculture's agenda. The advance of spatial control could only have occurred with the onset of modernity; as the technology developed to map and police kingdoms and ultimately nations, and the spaces therein, vaguely defined borderlands became sharply defined borderlines. The coves of pirates, the shires of highwaymen, peripheral places from where Kronos could challenge Zeus, were eventually walled and watched by the Cyclopes and Hekatonkheires. Some examples from the ancient, medieval, and modern world show the evolution of spatial control for the purpose of maintaining order through teleological stasis. The political institutions of ostracism and exile in the ancient world and the dispersal of urban crowds in the medieval and the early modern world provide evidence with which to examine this ideological framework of spatial order and temporal chaos.

In the ancient world, a resonant means of punishment for someone seeking to challenge or undo oligarchical authority was banishment. This casting out, the dispersal of the threat from the core to the periphery, was effective because the periphery was unknown and threatening. The threat of banishment in ancient, Greek city-states of Athens, Megra, Miletos, Argos, and Syracuse, a legal procedure known as *ostrakismos*, not only meant the removal of an individual from the place which provided them with a sense of identity and purpose, but it also meant that the Gods of the city-state were no longer watching over them, as both identity and religion were intensely bound to the idea of homeland.[14] According to Aristotle, the practice of ostracism was introduced by Cleisthenes in the late sixth century BC. The spatial removal of a threat (often done preemptively) from the center to the periphery was used as a political weapon among those oligarchs vying for power among the *demos*. Plutarch accounts for the practice of ostracism in Aristedes: "Now the sentence of ostracism was not a chastisement of base practices, nay, it was speciously called a humbling and docking of oppressive prestige and power; but it was really a merciful exorcism of the spirit of jealous hate, which thus vented its malignant desire to injure, not in some irreparable evil, but in a mere change of residence for ten years." An ostracized man, while their "dignity and pre-eminence" were curtailed, was "banished for ten years, with the right to enjoy the income from his property."[15] When the process of ostracism, intended to be a check on oligarchical power, became an instrument to consolidate it, it was promptly abandoned. Again, from Plutarch:

> And when ignoble men of the baser sort came to be subjected to this penalty, it ceased to be inflicted at all, and Hyperbolus was the last to be thus ostracized. It is said that Hyperbolus was ostracized for the following reason. Alcibiades and Nicias had the greatest power in the state,

and were at odds. Accordingly, when the people were about to exercise the ostracism, and were clearly going to vote against one or the other of these two men, they came to terms with one another, united their opposing factions, and effected the ostracism of Hyperbolus. The people were incensed at this for they felt that the institution had been insulted and abused, and so they abandoned it utterly and put an end to it.[16]

In Rome, however, the practice of exile differed from the Greek practice of ostracism in that it was a voluntary option in the face of legal punishment. Once a citizen opted for exile, a decree of *aquae et ignis* (fire and water) barred the exile from ever returning to his homeland.[17] The exiled citizen lost his political rights; patricians viewed the loss of civic identity equal to the loss of their life. Under the Roman emperor Tiberius, exile became the predominant form of punishment for traitors against imperial authority. Here again, threats to the hierarchical order of the center were dispersed to the periphery; poets and philosophers accused of subversion and political support of senatorial opponents of the emperor were exiled. Cicero, Rufus, Ovid, Seneca, Plautus, and Epictetus were all sent out and beyond.[18] Both Greek ostracism and Roman exile put those subjected to its implementation and enforcement "outside" of the law.

The physically participatory nature of ancient political practices and the specific spaces in which they manifested made bodily presence necessary for involvement. The removal of a citizen from the political center of the *patriae* reduced their power and stripped them of their identity, which was intrinsically bound to a particularly defined sense of place. The enforcement of hierarchical order is spatially designated. In the ancient world, displacement was a mechanism of enforcement, as removing people beyond the parameters of power was a relatively simple process. City-states were not homogenous, nor were they particularly immense. While Rome's territorial empire was expansive, it was still possible to remove a threat to the periphery where client states were obligated to receive an ostracized citizen and usually did so without complaint. Once the threat was removed from the center, it was no longer a threat. Order was restored, the challenge to authority neutralized by its physical relegation.

In the Middle Ages, the spatial dispersal of threats to maintain hierarchically imposed order was common procedure. The Peasants' Revolt of 1381, which threatened the reign of the fourteen-year-old Richard II and the landowning classes that were acting through his authority to collect extortionate taxes and suppress wages, was met at its climax in the political center of London with warnings to disperse. When a mob of dispossessed commoners had quite violently made their way to the hospital of St. Katherine's, adjacent to the Tower of London where the boy king resided, the unknown author of the *Anonimalle Chronicle* explains that the Richard II "had it proclaimed that they should go peaceably to their homes, and he would pardon them all their different offences."[19] When the crowd refused, demanding "charters to free them from all forms of serfdom," the king granted the requests on the condition "that every one should now quickly return to his own home." The crowd agreed, only to reassemble to hear the result of Richard's diplomacy, which they deemed unsatisfactory. Further negotiations with the crowd that had descended on London seeking redress from injustice were deadlocked, prompting the king to ask Wat Tyler, one of the revolt's leaders, "why will you not go back to your own country?" Tyler demanded, among other things, "that henceforward there should be no outlawry (ughtelarie) in any process of law, and that no lord should have lordship in future, but it should be divided among all men . . ."[20] The King agreed to Tyler's demands, only to have him killed while a contingent of armed men surrounded the crowd gathered outside of the Tower of London.

The death of Wat Tyler signaled the breakdown of negotiations brought an end to the peasant occupation of London as 20,000 conceded defeat and departed London on orders of the King. Once dispersed, Richard II abrogated the charter he had earlier drawn up to address the grievances of the peasants on the grounds that it was written under duress.[21] After their removal from the center of power, monarchical authority was reestablished and asserted with punitive consequence. Similar outbreaks in the east, north, and west were met with overwhelming force, and Parliament enacted punitive treason laws which resulted in the summary executing of hundreds of commoners.[22]

A primary reason for the Peasants' Revolt of 1381 was their systematic dispossession from the rural commons. A new system of land lordship and tenancy which relied on closing off access to land, and the resources it provided, resulted in destitution and hardship for rural people who had depended on the commons as an economic, social, and political resource for generations. This process, called "enclosures," was a major reason for the mob's descent into London that year. Henry Knighton's chronicle of the event recorded that "the rebels petitioned the king that all preserves of water, parks and woods should be made common to all: so that throughout the kingdom the poor as well as the rich should be free to take game in water, fish ponds, woods and forests as well as hunt hares in the fields—and to do these and many other things without impediment."[23] The enclosure of common lands, and the accessible wealth they provided, was one of the main features that drove the dispossessed into cities during the incremental shift from the medieval to early modern world from the fifteenth through the seventeenth centuries. Whereas with the Peasant Rebellion of 1381 the dispossessed rustics descended upon the city of London to issue demands only to be dispersed back to the countryside where they were subsequently "put down" by sword and law, the urbanization of the modernizing world presented hierarchical authority with new challenges in the control of space and the people who occupied it.

As the shift from a feudal to an industrial economy occurred with the incremental onset of modernity over the course of the sixteenth through the eighteenth centuries, the spatial dispersal of threats from the core to the periphery became an increasingly less effective means of maintaining hierarchical order. The lesson of John Milton's epic poem, *Paradise Lost*, published in 1667 in the aftermath of the English Civil Wars, was that simply casting rebellious factions out was no longer an effective means of hierarchical control. The story of Civil War between God and Satan begins *in media res*, with Satan and his revolutionary army having been banished to Hell, which Milton also refers to as Tartarus in the poem. It is from the depths of Hell that Satan, along with his generals Mammon, Belial, and Moloch, plot their revenge. Rather than renew direct conflict with God and his army in heaven, Satan opts to pursue a proxy guerrilla campaign by seeking to wreak havoc on God's newly created material world and turning humanity against him. Milton gave voice to Satan's plan:

> There is a place
> (if ancient and prophetic fame in Heav'n
> Err not) another World, the happy seat
> Of some new Race call'd Man . . .
> . . . this place may lie expos'd
> The utmost border of his Kingdom, left
> To their defense who hold it: here perhaps
> Some advantageous act may be achiev'd
> By sudden onset, either with Hell Fire
> To waste his whole creation, or possess
> All as our own, and drive as we were driven,
> The puny habitants, or if not drive,
> Seduce them to our party, that thir God
> May prove thir foe . . .[24]

God's wrath in reaction to this series of events is biblical, but the poem serves as a lens with which to view the period in which it was written as well as its religious subject matter. Direct conflict with authority in the spatial centers of power within the temporal context of nascent modernity was perilous; both Belial and Mammon advised Satan against it. When in pursuit of a teleological path that sought to undo the very system its architects were implementing, it was better to strike, indirectly, from the periphery, in order to upset the hierarchical and exploitative order of things. Only from a distance could the beast strike at the sovereign.

And so the beast acted against the sovereign from the periphery for much of the early modern period. Highwaymen, pirates, and bandits all challenged the emergent order of hierarchical control on its outskirts, just beyond the grasp of officialdom. As Eric Hobsbawm suggests, "banditry is a rather primitive form of organized social protest."[25] The rise of piracy and banditry from the sixteenth through the nineteenth centuries parallels the rise of merchant capital during the same period. As a result, the capitalist classes put the processes of modernity to work, triangulating the borderlands through mapping and surveillance, effectively smoking out the challenge from below in an attempt to establish the hierarchical ordering of modernity as an omnipresent entity that is everywhere all the time.

One of the first highwaymen of record who challenged the appropriation of the commons, the consolidation of wealth, and the unequal justice it produced is Robin Hood. While the debate about whether Robin Hood is legend based on an actual person or even an amalgamation of several, or a myth based solely on an idea or concept is dynamic; the fact that the legend has continued on to this day in such heroic form suggests something about the way we perceive of the systems that Robin Hood and his band of outlaws stood in opposition to. How many children grow up rooting for the Sheriff of Nottingham to put the lawless Robin Hood to the noose? Not many, if any at all. Yet, most adults wind up working for the same authorities that the Sheriff of Nottingham served in some way or another as so many agents of the technocratic monoculture just doing their job. Robin Hood and his Merry Men are remarkable for their ability to stay outside of the system that sought to incarcerate him, both physically and ideologically. The idea of Robin Hood challenging authority from the cover of the forest, away from the official center of authority, is romantic in our world of triangulated space and constant surveillance of people and places. Still, Robin Hood resonates in the modern world. Stephen Knight accounts for five Robin Hood films having been made before 1914, long before "The Golden Age" of movies.[26] At a time when modernity was evolving in new places with new technologies, Robin Hood was an archetype of resistance to emergent order that was both inspirational and informative. The legend continues to evolve in ways that reflect not only our interest of the past, but also how we perceive of the present. As children, we all root for Robin Hood. As adults, most eventually leave the shire and go to work for the Sherriff of Nottingham. It is easier to concede to hierarchical authority than to resist it, and so we leave Robin Hood by the bedside for our children to read and to dream, while we prepare for work the next day.

At sea, pirates stood in defiance to the emergent order of capital accumulation since ancient times. The acceleration of capital accumulation in the early modern world also increased the amount of piracy. The early part of the eighteenth century is generally regarded as "The Golden Age" of piracy, for the sheer scale of the operation as well as its efficacy. Marcus Rediker observes the many ways that pirates stood in defiance to the emergent conditions of modernity: their crews were international in scope in an age of emergent nationalism, while the pirates themselves "inhabited a world huge, boundless and international." They worked cooperatively for the benefit of the whole when much of the labor force was being organized to work for the individual benefit of a few. Pirates challenged the hierarchical authority of the ship, and by proxy, the political economy of which it was a part, by engaging in direct democracy and collective decision making. They reappropriated the wealth that had been appropriated from the commons at the onset of early modern accumulation by seizing ships which traded in raw materials and finished goods. Pirates, in sum, were part of a larger cycle of resistance against the forces which "sought to make 'authority and obedience' the 'rule of the land'" and left an example from which later radicals "who continued the fight for democracy and freedom" could learn.[27]

Banditry, outlawry, and piracy are traditional challenges to the political economy of command and control that continue today. In the Gulf of Aden, Somali Pirates have interrupted international shipping lanes since the earliest part of the twenty-first century. As a result of environmental degradation due to illegal toxic waste dumping and the encroachment of foreign commercial fishing vessels illegally operating in the United Nations designated Exclusive Economic Zone in Somali waters, fishermen who had plied their trade in a traditional manner were unable to continue doing so.[28] The turn to piracy was a result of the degradation of the commons, their pursuit of trade vessels a reaction to the theft of their livelihood. As a result of their challenge to the wholesale theft and destruction of their coastal waters, the international community has increased its military naval presence in the Gulf of Aden. Triangulating and policing the spaces in which the disenfranchised challenge the political economy with martial force, rather than finding ways of minimizing its impact on people who live outside of it, is the manner in which the global economy enforces technocratic monoculture.

Piracy also exists in virtual space. Internet piracy, the trading in music and video files, is a fairly common phenomenon. The commonwealth of file sharing is something that the music and film industries have sought to crush as so many eighteenth-century autocrats hunted the pirate ships of the Atlantic. Indeed, one major file sharing network is called "The Pirate Bay." While part of the efficacy of ending piracy in the eighteenth century was the brutal and public execution of pirates as a means of terrifying others who would sail under the black flag, the modern entertainment industry has engaged in the public punishment of a few transgressors to make examples of them and discourage others from sharing what they should be purchasing from corporate sources. In 2012, the eighth U.S. Circuit Court of Appeals reinstated a verdict against Jammie Thomas-Rasset, ordering her to pay $222,000 to the Recording Industry of America, a corporate lobbying group, for downloading twenty-four songs. The example the RIAA set in pursuing and ruining Thomas-Rassett rings of the eighteenth-century executions of pirates by hanging. The corpses of pirates were often left to hang and rot at the docks as a warning to others who would challenge the authority of capital and state at sea.

Similarly, present-day outlaws inspire as much fascination as did their early modern counterparts. Ralph "Bucky" Phillips was imprisoned at the Erie County Correctional Facility in Alden, New York in April of 2006 for violating his parole terms after a burglary conviction (he escaped days before his official release date). Philips was able to avoid capture until September;

for fully five months, he evaded the panoptic eye of the law enforcement community. Like Robin Hood, Phillips is believed to have spent most of his time as a fugitive in the forests of Western New York and Northern Pennsylvania, and during his time on the run was responsible for the death of one law enforcement officer. Phillips's story also has parallels to the story of Jack Sheppard, the early eighteenth century gaol breaker of London, who inspired a great degree of excitement for his ability to evade detection. Like Sheppard, Phillips became something of a folk hero in Western New York. A cottage industry of T-shirts and fast food items bearing his name and likeness popped up: "Run Bucky Run" became a familiar refrain.[29] The interest in Philips prompted New York State police to complain about public sympathy for Phillips in the press coverage surrounding the manhunt. Daniel M. DeFedericis, president of the Police Benevolent Association, insisted that "the only heroes in this case were the troopers who risked their lives and were shot. The villains are Phillips, and to a lesser degree, the people who support him."[30] Philips was eventually caught, tried, and convicted of murder. When asked how he pled in court, Philips replied: "Guilty as hell."[31]

With the growth of cities in the early modern industrial era, disturbances and tumults within cities became more common as the processes of modernity systematically dispossessed people of political rights and material wealth. The late seventeenth and early eighteenth centuries saw a proliferation of civil disturbance in London, the axis of the modernizing world, as industrialization was taking root across England and the assertion of authority over civil space became more comprehensive. The Bawdy House Riots of 1668, the Sacheverell riots of 1710, the Coronation Riots of 1714, and the 1715 Riots all threatened London's powered elites with a spatial proximity previously unknown before the predominance of the urban condition. To deal with this new and threatening phenomenon, and to protect the institutions of hierarchical authority, Parliament passed the Riot Act, which took effect in 1715. Formally titled "An Act for preventing tumults and riotous assemblies, and for the more speedy and effectual punishing of the rioters," this act sanctioned the declaration by authorities that any group of twelve or more people unlawfully assembled would have the choice of either dispersing after one hour of being noticed or "suffer death."[32] The Riot Act was an antiquated idea, suitable for a time when crowds could be dispersed back to whence they came, usually outside of the bounds of the city. When the rioting crowds lived within the city limits, the authorities were essentially justifying bloodshed. Rather than listen to the demands of the crowd, the defenders of status quo order provided a legal framework so that anyone participating in such civil assembly "shall happen to be so killed, maimed or hurt."[33]

Maintaining civil oppression by preventing people of like mind to gather in hopes of seeking redress became a primary weapon in the arsenal of hierarchical control in the modern age. Just as the dispossessed urban poor in London were warned from assembly under pain of death after a series of riots that threatened the hierarchical order of nascent modernity, acts passed in the wake of slave rebellions in the United States in the eighteenth and nineteenth centuries sought to prevent politically, economically, and socially disenfranchised slaves from gathering to discuss and act upon their situation. Both in the urban centers of the eastern seaboard and in the rural plantation communities of the antebellum South, where the institutional elite depended on slave labor as the foundation of their entire economic system, laws to control the slave population by controlling their actions and behaviors in spatial terms proliferated. In fact, laws prescribing the comings and goings of slaves in North America proliferated around the time that tumults which led to the passage of the Riot Act in England were occurring. For example, the South Carolina General Assembly passed a statute entitled "An Act for the Better Ordering of Slaves" in 1690, which expressly sought to enforce and maintain the tenuous hierarchical order of the nascent slaveocracy. The Stono Rebellion of 1739, in which nearly fifty whites were killed by a rebellious faction of slaves in one afternoon, resulted in the passage of the Negro Act of 1740, which, among other restrictions, made it illegal for slaves to assemble in groups. Using language similar to the Riot Act of 1715, it called for authorities "to disperse any assembly or meeting of slaves which may disturb the peace or endanger the safety of his Majesty's subjects."[34] Similar acts were passed throughout the North American colonies to maintain hierarchical order by preventing the chaos that crowds of dispossessed persons could cause.

From the ancient practice of ostracism and exile in Greece and Rome to prevent one oligarch from gaining too much power, to the dispersal of crowds in the early modern era to protect oligarchs from having to concede any power, the element of spatial control to prevent change in social, economic, and political condition remains constant. The reasons for the shift from spatial relegation as a means of maintaining balance to consolidating it lie within the context of modernization. As the development of modern market economies was based on the inherent inequalities of class and race, dispersing crowds who sought to challenge the institution of this system was possible due to the largely rural condition of the subject population. With the rise of urbanization as a central feature of modernity, new means of controlling space became necessary, given the proximity of the disenfranchised to the institutional elite. Once peripheral, they were now central. With the territorial expansion of national authority in modernity, there was no more periphery toward which authority could relegate and marginalize the threat to institutional order. Once the crowd could no longer be dispersed, the Hekatonkheires built their walls and the Cyclopes stood guard. Controlling crowds became the next step in the battle between order and chaos. New technologies to monitor, police and control the activity of the modern dispossessed became necessary. Zeus unleashed his Cyclopes and Hekatonkheires in a final bid to unseat Kronos and establish his control over Olympus.

MONUMENTS AS SYMBOLS OF TEMPORAL, SPATIAL, AND IDEOLOGICAL STASIS

While spectacular events resonate on the historical record as indicators of fundamental changes that shape our present condition, monuments serve the paradoxical function of commemorating and memorializing those changes in ways that fix them as permanent, both in time and in space. While not particularly modern phenomena, monuments have proliferated in the modern epoch in a manner which evokes Moore's Law of Accelerating Returns, that is to say, exponentially so. The American theologian Reinhold Neibuhr observed in his essay *The Tower of Babel* that "one of the most pathetic aspects of human history is that every civilization expresses itself most pretentiously, compounds its partial and universal values most convincingly, and claims immortality for its finite existence at the very moment when the decay that leads to death has already begun."[35] The proliferation of monumental design in modern architecture exposes the technocratic monocultural values of modernity at a moment when the zeitgeist is preoccupied with what many believe to be its inevitable collapse.

Monuments operate as a means to shape, frame, or otherwise control the discursive memory of a "spectacular event" in a way that serves the interests of the power structure commissioning the memorial. By reifying an event or person in spatial and material terms, monuments objectify the meaning of their subjects in ways that often reduce and simplify them into forms that reinforce and add to the official narrative already in place. As Sanford Levinson suggests in *Written in Stone: Public Monuments in Changing Societies*, ". . . those with political power within a given society organize public space to convey (and thus to teach the public) desired political lessons."[36] The institution of public memory as a creation of the state and the marketplace creates icons, and challenges to the sanctioned narrative are tantamount to iconoclasm. The ideological borderline of acceptable discourse is drawn across piety and heresy.

Monumental space renders public memory passive and deferential. The viewer is left separated from the events and the actors that monuments commemorate. Here, the monument serves the dual purpose of transmitting a sanctioned historical narrative and imparting identity onto the viewer. At once, the viewer is a patriot and consumer, and their understanding of both official and commercial memory becomes vernacular in the process. "The national myths," as Katharine Hodgkin and Susannah Radstone demonstrate in *Contested Pasts: The Politics of Memory*, "are driven by a desire either to conceal, or at least to assign safely to the past, the marks of injustice and inequality, of tensions between oppressors and oppressed."[37] In this process, the commodification of memory is complete. The viewer does not actively participate in the narrative, they stand awestruck and deferential in the space designed to reinforce an ideological message. Monuments encode the power of place in political, economic, and cultural terms. The time they commemorate is distanced, separated, and isolated from the present. Revolutions *happened*. Great men and women *challenged* the status quo. An event of particular significance *occurred* here. The point was not lost on Henri LeFebvre, who wrote that:

> The monument thus affected a 'consensus,' and this in the strongest sense of the term, rendering it practical and concrete. The element of repression in it and the element of exaltation could scarcely be disentangled; or perhaps it would be more accurate to say that the repressive element was metamorphosed into exaltation.[38]

Any challenge to the hegemonic message that separates past revolutionary disorder from present repressive order is "out of place" in monumental public space. Attempts to connect the past and present in monumental space are criminalized. In the wake of a fatal police shooting in Athens in 2008, Greek protestors hung banners from the Acropolis calling for protests throughout Europe on the 18th of December in that year. The police shooting compounded tensions in Greece; the tensions were the result of poor economic conditions and political corruption. An anonymous protester told reporters that "we chose this monument to democracy, this global monument, to proclaim our resistance to state violence and demand rights in education and work. (We did it) to send a message globally and to all Europe." In response, government spokesman Evangelos Antonaros said that "there can be no justification for this action. This hurts the image of our country abroad . . . it is unacceptable."[39] While the banners were taken down after two hours, riots in Greece lasted for three weeks, with solidarity demonstrations taking place in more than seventy cities around the world.[40]

Unauthorized participation in the narrative frame of public monuments in a way that challenges the account can be criminalized as vandalism, even if the alterations are temporary and removable. In 1989, American Indian Movement activist Russell Means was arrested in Colorado for pouring blood on a statue of Christopher Columbus on Columbus Day, which officially commemorates his "discovery of America" and recognizes Columbus as a "pioneer of progress and enlightenment." *The Rocky Mountain News* ran an editorial dismissing his protest as "petty vandalism," and asserted that "pouring gunk all over public monuments goes beyond speaking or burning your own flag and into the kind of behavior that should be rewarded with some mandatory community service."[41] By comparison, the alteration of monumental statuary to reinforce a political or commercial narrative is entirely sanctioned and therefore acceptable. The frequent dressing of civic statuary with sports jerseys is organized by the city government, for the purposeful promotion of the corporatized interests of those teams, in hopes that the ancillary civic pride that comes along with their success resonates in the public sphere. Here, the agenda of public monuments as spheres of spatial and ideological bolstering for hierarchical capitalist order becomes readily apparent.

The architects of monumental structures and spaces employ both abstract and literal designs in their attempts to commemorate and reinforce their intended narratives. The pyramids of Giza are not literal representations of their respective Pharaohs, but they symbolize the power that the Pharaohs held in their lifetimes. More literal monumental commemorations, for example, Roman imperial statuary, attempt to represent the body and figure of powerful patrons in often larger-than-life form. This abstract/literal dynamic is something that carries on to the present day. The Washington Monument in Washington, D.C., an obelisk that represents George Washington's primacy as a national leader, is decidedly abstract, whereas the Lincoln Memorial is a more literal representation of the sixteenth President of the United States, albeit on a purposefully exaggerated scale. The exaggerated spatial architectonics of monuments resonates in the temporal sense as well. LeFebvre explains that "the most beautiful monuments are imposing in their durability. A cyclopean wall achieves monumental beauty because it seems eternal, because it seems to have escaped time."[42] In the case of the Washington Monument and the Lincoln Memorial, as with most (but not all) monumental memorials, the viewers' eyes are drawn upward, to appreciate the power and greatness of the subject in memoriam. As Yi-Fu Tuan explains, height and status are synonymous in modern spatial reckoning.[43] The taller the statue, the more idealized the representation is, the less it is bound in spatial and temporal terms. Similarly, the higher up a viewer has to look to view a statue, the less powerful their position is in relation to the subject of the statuary.[44] Our temporal mortality becomes subtext to the immortality of the monumental structure and subject. These architectonic elements all contribute to the structuring of monumental space and ideology, which reinforces technocratic monoculture, interrupts teleological change, and imposes ontological stasis.

A spectacle is a fixture in time that is oriented in space. A monument is a fixture in space that is oriented in time. Within the modern technocratic monoculture, the purpose of monumental construction and design is to appropriate the memory of events that advance the cause of teleological dynamism and render them static, in service of the very institutions their occurrence challenged. Debord explained the rupture and reconfiguration accordingly: "Time remains motionless, like an enclosed space. When a more complex society finally becomes conscious of time, it tries to negate it."[45] Monumental space renders the past and future separate from the present, ossifying teleology as a means of empowering the hierarchical interests that seek to impose and maintain ontological stasis. As Maya Nadkarni describes it, "the monument effaces its own existence in the present because its hypostasis and temporal completion negate any potential for future transformation."[46] Power, presented in monumental form, assumes temporal authority over the impetus and direction of political, economic, and social change in a manner that reinforces its interests to altogether either prevent its occurrence, or render it in superficial terms. This is done in order to promote a discourse of change to mitigate or otherwise deflect a popular awareness of the repressive nature of imposed and enforced ontological stasis.

Rather than engage an inventory of monuments both ancient and modern here, which has been done much more thoroughly with altogether different purposes by other scholars, a few serve as representative of larger themes regarding the spatial control and management of temporal events within the context of modernity. The Boston Massacre and Tyburn Memorials share common themes of simultaneously commemorating and concealing the historical agency of the lower classes in the transatlantic world of the seventeenth and eighteenth centuries. Fort Meigs in Perrysburg, Ohio constructs a triumphant national narrative while marginalizing the native population that lived in the region prior to national expansion. The Haymarket memorials in Chicago show how contested narratives can be buried by abstract relativism in discourse and design. Lastly, the 9/11 Memorial obfuscates questions about global imperialism and what the nation represents, and instead reinforces a nationalist narrative through commemoration.

LOOKING DOWN: THE BOSTON MASSACRE AND TYBURN MEMORIALS

Some monuments do the seemingly paradoxical job of attempting to diminish the resonance of the thing they are designed to commemorate. They are begrudging acknowledgements of past injustice that the state would prefer to minimize in order to maintain the authority that the events they commemorate call into question. The Boston Massacre and the Tyburn Tree in London are two examples that serve to illustrate this point. At the heart of the Boston Massacre lies a street protest of economically and disenfranchised members of the underclasses. Tyburn is the ultimate manifestation of "death from above," a state-sanctioned spectacle of murder to terrorize the population and subject it to legal and commercial authority. Both monuments commemorating these histories are designed to minimize the role of common people in the events they represent, and understate the role that the spectacle of violence has in subjecting citizens to the authority of the state.

The Boston Massacre, or as it was known in Boston before its propagandization by Paul Revere, the Riot on King Street, was a temporal event that happened in a particular place. On March 5, 1770, underemployed Bostonian dockworkers clashed with British regulars who were taking second jobs at the docks for lower wages to supplement their inadequate incomes from the crown. It wasn't so much a political event as it was a labor dispute.[47] The clash resulted in the death of five colonists: Crispus Attucks, Samuel Gray, James Caldwell, Samuel Maverick, and Patrick Carr, three of whom died on the spot when British soldiers discharged their weapons into a crowd that had assembled to settle a fight from a previous encounter at the Rope Walks a few nights earlier. Six others were wounded in the melee.[48] The event was used by colonial elites to rally the cause of Revolution throughout the thirteen colonies, and bring them toward a common identity that would help define what would eventually become American national identity.

In contrast, the Tyburn Tree was a particular place where numerous temporal events occurred. The first hanging at Tyburn occurred in 1196: the execution of William Fitz Osbert for the crime of advocating for the poor and leading them in a revolt against the concentrated wealth of London's elites. As William of Newburgh put it in his 1196 chronicle *Historia rerum Anglicarum*: "At length, by his secret labors and poisoned whispers, he revealed, in its blackest colors to the common people, the insolence of the rich men and nobles by whom they were unworthily treated," and "took upon himself to plead the cause of the poor citizens against the insolence of the rich" so that ultimately "a powerful conspiracy was therefore organized in London, by the envy of the poor against the insolence of the powerful."[49] For the crime of observing injustice, Osbert was strung up on a tree near a stream. In 1220, the tree was replaced with a pair of gallows, and in 1571, as public executions in London were reaching their apex, the famed "triple tree" was erected. It stood as a three dimensional triangle eighteen feet high with three equilateral nine-foot beams to accommodate the execution of up to twenty-four people. The dimensionality of the gallows was a principal feature of the spectacle that occurred there every Monday until 1783, when the executions were moved to Newgate Prison to minimize the role of the crowds that gathered to bear witness to the spectacle of state power over its subjects. Poets used Tyburn as a point of reflection and London guidebooks used Tyburn as a landmark to orient tourists.

Today, both memorials at the places where the Boston Massacre occurred and where the Tyburn Triple Tree stood are absent of any dimensionality. The site of the Boston Massacre was commemorated in 1887 with a cobblestone circle in a traffic island at the corner of Devonshire and State (formerly King) streets, in front of the Old State House.[50] Much of the funding was raised by the African American community in Boston in an attempt to tie African American freedom to the cause of nationalist patriotism given that Crispus Attucks was an African American and largely regarded as the first "martyr" of the American Revolution. The star at the center of the cobblestones was intended to represent Crispus Attucks, though neither he nor any of the other victims are explicitly named on the marker.[51] In 2009, it was moved, somewhat haphazardly (where exactly the bodies fell on March 5, 1770 is a matter of debate), to the sidewalk in front of the State House in an effort to minimize the danger of tourists venturing out into the pedestrian island through the busy Boston traffic. The marker was also given brass accents to surround the cobblestones and ringed with the words "Site of the Boston Massacre March 5 1770" (See Figure 4.2). A larger monument, erected in 1888, stands in Boston Common, but it is deracinated from the place of the event. Its construction was protested on ideological grounds by the Massachusetts Historical Society, who argued, in a legal suit to stop it,

Figure 4.2 Boston Massacre Site Memorial, 2014.
Photo by Brian Frawley.

that "the martyr's crown is placed upon the brow of the vulgar ruffian."⁵² Its arbitrary location relegates it as ancillary and removed from the historicized environment of the Old State House, at the same time embedding the actions of the massacre victims as coincidental to the cause of nationalist discourse. The iconography of lady liberty and a bald eagle stand at the foreground of the statue, behind this is a pillar on which the names of the five victims of the Riot on King Street are engraved. As Karsten Fitz observed, such representations were part of a larger trend in the 1880s, which accounted for "the erasure, the marginalization" and the "appropriated, passive" reemergence of Attucks as "an African-American patriot, within—and not outside or against—the overarching national narrative."⁵³

As the height of a monument is historically related to the importance of the subject it is commemorating, and has been since the Egyptian pyramids were built, the lack of height is the first factor worth noting in the design of both memorials. Both were at one point located in traffic islands (as previously mentioned, the Boston Massacre "site" has been moved to the sidewalk for safety reasons) as the commercial space surrounding the sites upon which both exist was ultimately deemed too valuable to preserve for memorial purposes. Rather than illustrate the actual nature of the street-level conflict that led to the massacre, the city of Boston prefers to perpetuate Paul Revere's genteel representation of the events, which portray the rioters as passive white gentlemen victimized by a coordinated English attack. A

Figure 4.3 Tyburn Memorial.
Photograph by author.

reproduction of Revere's famous woodcut greets commuters as they exit the subway stop below the Old State House.

The present Tyburn Memorial is a small concrete circle in a traffic island on Edgware Road in North London, with a small, seemingly arbitrary Maltese cross surrounded by the inscription "The Site of Tyburn Tree" (See Figure 4.3). It is a replacement for the original, which was "a six foot triangle with an inscription in brass letters set in granolithic: 'Here stood-Tyburn Tree-Removed 1759,'" installed in 1909 by the London City Council.[54] The current replacement was put in by the London County Council in September 1964 without much fanfare or official record; the *Glasgow Herald* observed the "equally lethal" location of the memorial in a traffic island, and that the memorial "discreetly commemorated" Tyburn with "a clinical circular slab of Portland stone."[55] As Alan Brooke and David Brandon note in their book *Tyburn: London's Fatal Tree*, it is "a location that does not encourage quiet contemplation of the sights and sounds of the past."[56] Much of the commemoration of Tyburn that rings the traffic island is unofficial. Tyburn Convent acknowledges the Catholic martyrs who hung at Tyburn, and a public house (run by the J. D. Wetherspoon pub chain) called "The Tyburn" on Edgware Road superficially acknowledges on its website that "this Wetherspoon pub takes its name from Britain's most famous place of execution."[57]

The Lloyds Bank on Edgware Road keeps a stone post inscribed with the words "Tyburn Gate" in its window, along with a plaque acknowledging the history of Tyburn as a place of execution. In 2009, plans were seemingly underway to build a monument to specifically recognize the 105 Catholic martyrs who died at Tyburn, though estimates place the total number of executions at Tyburn between 40,000 to 60,000.[58] The City of Westminster acknowledges the existence of the memorial plaque on the website for the Marble Arch. Last updated in 2010, the last sentence of the site suggests that "there are also long-term plans for a larger more appropriate memorial within the marble arch site."[59] As of this work, the small cracked circle remains, while the Marble Arch, a remnant of a since-replaced ceremonial entrance to Buckingham Palace, stands nearly three hundred feet high across the street at the Northeast Entrance to Hyde Park and casts a direct shadow over the broken concrete memorial at certain times of the day.

Monuments, in recalling spectacular events that shaped the course of history, serve to reify and reinforce control of the narrative discourse in spatial and ideological terms. This appropriation of memory and meaning undermines the advance of ideas by distancing the past from the present through symbolism, which ultimately serves to enforce a static hegemony of order and power by putting the issues behind us rather than in front of us, and making them ambiguous enough to ignore altogether. Spectacles provide opportunities for further consideration of the pressing issues of modernity. Monuments render those spectacles mute and impose a teleological stasis to prevent revolutionary ideas from manifesting.

INVISIBILITY: FORT MEIGS

Whereas Tyburn was the end for many who resisted the hegemony of property rights in a nascent modern capitalist economy, and the Boston Massacre sparked a conflict over English hegemony in the political affairs of the North American colonies, the War of 1812 was a tripartite conflict. While the English looked to retain their influence in the Northwest Territories after formally ceding the territory to the United States in 1783, the Americans sought to consolidate their territorial holdings west of the Appalachian Mountains to make them economically productive. A third group, the Native American population who had lived on the land long before either the English or Americans asserted legal claims to it, vied for a degree of territorial and cultural independence and autonomy in the conflict. In a complex pattern of accommodation and resistance, a broad cross section of Algonquin and Iroquois nations sought to navigate through the imperial and national hegemony that the English and Americans were imposing on the territory.

The Native American leader Tecumseh figured prominently in the resistance, organizing native populations in the contested territories from the

Great Lakes and upper Mississippi regions and as far south as Alabama with a blend of nativist rhetoric and militant calls for resistance to further Anglo-American encroachment.[60] The defeat of the French in the Seven Years War was devastating to the native populations west of the Appalachians, as neither the British nor the Americans who successively asserted claims and rights to the land through force had plans to accommodate native intentions to stay on the territory. Indiana Governor William Henry Harrison invoked Hesiod's construction of mythological history and "The Golden Age" in a letter to the Secretary of War William Eustis regarding the Indians' perception of the situation:

> The happiness they enjoyed from their intercourse with the french is their perpetual theme—it is their golden age. Those who are old enough to remember it, speak with rapture, and the young ones are taught to venerate it as the Ancients did the reign of Saturn . . . said an old Indian chief to me . . . 'they never took from us our lands, indeed they were in common with us—they planted where they pleased and they cut wood where they pleased and so did we—but now if a poor Indian attempts to take a little bark from a tree to cover him from the rain, up comes a white man and threatens to shoot him, claiming the tree as his own.'[61]

From this perspective, native resistance can be read similarly to the resistance against the enclosures movement in early modern England. The idea of a transatlantic commons, and its appropriation by the nascent forces of technocratic monoculture, becomes clear and resonant here, and we can connect seemingly disparate groups through an ideology that sees the triangulation and commodification of space as an interruption to a universal teleological cycle. And like revolutionaries in the western world, the native population of North America articulated a hybridized eschatological and teleological prophecy that was at once temporally linear and circular:

> After these white people had landed, they were not content with having the knowledge which belonged to the Shawnees, but they usurped their land also. . . . But these things will soon end. The Master of Life is about to restore to the Shawnees their knowledge and their rights and he will trample the Long Knives under his feet.[62]

In the spring of 1813, Tecumseh led a band of natives in conjunction the British General Henry Procter to take the American Fort Meigs on the bank of the Maumee River in present day Perrysburg, Ohio. The raid ultimately failed, but in the aftermath of the battle natives, began to slaughter American prisoners who were caught outside of Fort Miamis, a few miles east, while English officers stood by. William G. Ewing, one of the American captives, said that he had heard "a thundering voice . . . in the Indian tongue." Tecumseh had arrived from the field to stop the slaughter. One English officer recounted that "his eyes shot fire. He was terrible." Tecumseh rebuked everyone in the camp, native and English alike: "Are there no men here?" When the natives drew away, Tecumseh turned to General Procter, who complained that "your Indians cannot be controlled; they cannot be commanded." Tecumseh replied "Begone! You are unfit to command. . . . I conquer to save, and you to murder."[63]

The incident contributed to Tecumseh's legend as a leader, which was largely recognized by his enemies only after his death. After their soldiers had mutilated his corpse, American generals and politicians lined up to sing his praises as a worthy adversary and a man of commendable character. General Lewis Cass, who fought against Tecumseh at the Battle of the Thames, Indian Agent John Johnston, and even William Henry Harrison praised his legacy. Harrison's praise framed Tecumseh within teleological and ontological conditions, calling him "one of those uncommon geniuses, who spring up occasionally to produce revolutions and overturn the established order of things."[64] In spite of these convenient constructions of noble savagery, Tecumseh's legacy on the landscape of American history remains relatively muted. Where his name does resonate, it is as mutilated as his corpse was in 1813. Five towns across the Midwestern United States somewhat ironically bear his name, but, according to the 2010 census, none have a Native American population above 1%. Even more ironically, four American naval vessels have borne the name *Tecumseh*. A company formerly based out of Tecumseh, Michigan called "Tecumseh Products," which makes air compressors, retains the name along with a logo of an Indian in profile with full feathered headdress, surely something Tecumseh never wore.

As monuments are typically meant to spatially commemorate an official account of a temporal event, attempts to challenge the hegemonic narrative are typically marginalized. At the Fort Meigs War of 1812 Battlefield in Perrysburg, Ohio, the largest reconstructed wooden-walled fort in the United States, the British and native role in the conflict is acknowledged, but largely in an adversarial context, which sets the ultimately triumphant "Legacy of Freedom" narrative that frames the experience in the newly constructed Visitor's Center. At the center of the fort itself stands an eighty-two-foot obelisk, dedicated in 1908 by The Grand Army of the Republic, "in recognition of the gallant men who defended their country on this spot." In 2008, those words on the base of the monument were painted over with the reminder "Tecumseh Was Here," alongside an anarchist symbol (See Figure 4.4). This unofficial reminder of the fact that the native population was also defending their country was literally whitewashed from the monument shortly thereafter. The official narrative and commemoration of the War of 1812 was thereby restored, and thus Tecumseh was removed from the place a second time.

Figure 4.4 Base of the Fort Meigs Memorial Obelisk, 2008.
Photograph by Daniel Wilkens.

THE MOVING TARGET: HAYMARKET

As monuments institute narratives that resonate over time, the discourse surrounding their establishment and meaning is often contested. While nearly every monument has such debates at its initial conceptualization, few monuments have been the subject of the prolonged politicized discourse as the Haymarket Memorial in Chicago, Illinois. The Haymarket affair, a catchall term representing the labor rallies, bombing, trials, and execution of four men in Chicago in 1886, was a quintessentially spectacular event, one that shaped the direction of the labor movement and reactions to it since that fateful night of May 4, 1886 when seven police officers and four civilians were wounded by a dynamite bomb. After the execution of four of the Haymarket defendants, George Engel, Adolph Fischer, Albert Parsons, and August Spies (a fifth, Louis Ling, committed suicide in his jail cell a day before his execution), a contest for the memory of what occurred at Haymarket began that continues to this day.

The first monument to commemorate the event was dedicated in 1889; it was a nine-foot statue of a police officer raising his right hand in a manner which is evocative of a fascist salute (See Figure 4.5). It was placed on a grandiose pedestal bearing the somewhat paradoxical inscription "I

Figure 4.5 Haymarket Police Monument.
Photograph by author.

Command Peace" directly at the center of Haymarket Square. The entire monument "of heroic size," over twenty feet tall in total, intentionally sent a message that law and order had reclaimed the place where labor activists had attempted to upend the socioeconomic order of things.⁶⁵ The statue was taken down in 1900 under pretense, as merchants were complaining that "it was in the way of their business and interfered with traffic."⁶⁶

Capitalism and technological progress were likely the only acceptable reasons to remove the statue, though it was clear that the statue was an imposition and an openly ominous reminder of the neighborhood's subordinate status as a base for Chicago's working-class, immigrant population. That the statue was occasionally vandalized showed that tensions from the 1886 incident had not abated. The statue of the officer, with its pedestal, was then moved to Union Park, where a streetcar driver jumped his tracks to hit it on the forty-first anniversary of the Haymarket Bombing in 1927. The driver complained that he was "sick of seeing that police officer with his arm raised."⁶⁷ From there, it was moved further into Union Park to prevent motorists from taking a second shot at it.

In 1958, with the Cold War in full swing, the statue was moved back to Randolph Street, just 200 feet from its original location. By then, a highway had been built, and a special platform was erected so the bronze law enforcement officer could overlook the traffic. Back from the periphery and returned to the center, the statue symbolized a return to the idea of law and order as enforced from above in 1950s America. Between October 1969 and 1970, at the height of social and political turbulence in the United States surrounding the Vietnam War and Civil Rights Movement, the statue was damaged twice by explosives and once by printer's ink. The Weathermen, a radical organization committed to the overthrow of the United States government, claimed responsibility for both the 1969 and 1970 bombings. After the second bombing, the Weathermen took credit by phoning police headquarters with this message: "We just blew up Haymarket Square Statue for the second year in a row to show our allegiance to our brothers in the New York prisons and our black brothers everywhere. This is another phase of our revolution to overthrow our racist and fascist society. Power to the People."[68] By orders of Mayor Daley, the statue received twenty-four-hour police protection until 1972, when it was taken from its pedestal and was moved to Police Headquarters at the corner of 11th and State Street. Still an object of resentment and ridicule, the statue was moved to the courtyard of the Police Academy in 1976, where it was inaccessible to the public until its rededication with a new pedestal at police headquarters in 2007. Meanwhile, the pedestal remained without its sentinel until its removal in 1996, the object of continued vandalism throughout.

In 1893, four years after the dedication of the "official" monument commemorating the Haymarket Bombing, a somewhat similar monument in form, but very different in tone was dedicated at Waldheim Cemetery in Forest Park Illinois (See Figure 4.6). The centerpiece is a sculpture of a woman standing in front of a dead man while her right hand clutches her breast in a stoically protective posture. At the foot of the monument supporting the work is chiseled a quote attributed to August Spies before his execution: "The day will come when our silence will be more powerful than the voices you are throttling today." The back of the monument lists the names of the Haymarket Martyrs, as they came to be known, and above their names hangs a plaque citing the pardon for Samuel Fielden, Michael Schwab, and Oscar Neebe that Illinois Governor John Atgeld issued the year the monument was dedicated.[69] Like the statue of the officer commanding peace, this statue demands justice. Both have symbolically waged their debate for over a century, but neither occupies the ground of the event in question. That territory was ceded until 2004, when a third memorial was built.

The third and most recent monument to commemorate the bombing at Haymarket Square, known as the Haymarket Memorial Sculpture, was the result of the joint efforts of the Illinois Labor History Society, several labor unions, state bureaucrats, and politicians (See Figure 4.7). The Memorial, sculpted by American artist Mary Brogger, is an abstraction of the speaker's cart on which labor activists stood to speak to the crowd on the fateful night of May 4, 1886. It stands on the very spot where the wagon is believed to have stood. Its spatial accuracy is betrayed, however, by the purposeful abstraction of the event itself in artistic and narrative terms. The result is an attempt to apply a postmodernist past-exonerative tense which denies agency to anyone

Figure 4.6 Haymarket Martyrs Memorial.
Photograph by author.

Figure 4.7 Haymarket Memorial.
Photograph by author.

involved in the events, in an attempt to somewhat politely concede that, in a Nixonian bureaucratic passive voice, "mistakes were made." Abstracted figures, faceless and anonymous, stand on and around the cart in a chaotic swirl in which they appear to either be constructing or deconstructing the wagon. Brogger purposefully left this open to interpretation, stating that "I was thinking about the nature of anarchy, how you have to destroy something to build something." She suggested, again using the passive voice, that the sculpture is ultimately "a cautionary tale about how power is wielded."[70] By whom, against whom, and for what purpose, are questions that are left open to interpretation by the ambiguous nature of the work. There is no resolution.

At the dedication ceremony in September 2004, Fraternal Order of Police President Mark Donahue declared that "organized labor and law enforcement have both come a long way in the past 118 years. Law enforcement is now part of the labor movement. The efforts of the pioneers of the labor movement and the sacrifices of those eight police officers who were killed have not gone unanswered. Organized labor has enjoyed and our country has reaped the rewards of those efforts."[71] Here, Donahue cleverly associated law enforcement with the very movement it tried to crush on behalf of the business interests it represented. Flanked by labor leaders, politicians, and bureaucrats, he was briefly hollered down by anarchists waving black flags at the ceremony, who complained that the institutional powers represented on the podium were "whitewashing" history, but the ceremony proceeded to conclusion.[72]

The intended symbolism of this dedication ceremony, both sides literally coming together over common ground—the anarchists and the authoritarians—and engaging in dialog and compromise, is arguably at the expense of the entire debate that, like the sculpture, has become deliberately abstract and subjective. In spite of the best efforts of everyone involved to bring closure to the Haymarket Affair and all it represents, there really is no comemory of what happened or what it means to us today. The police still have their statue, the anarchists still have theirs. Both are in separate corners, as it were, and the ground on which the original clash occurred has been rendered peaceful by ambiguity. The subjects of debate, class warfare and the oppressive violence of a police state supported by industrial capitalism, are altogether removed from consideration. On the very spot where Samuel Fielden warned the crowd that "the law is your enemy" while Adolph Fischer distributed flyers that urged "REVENGE" for the murder of locked out workers the previous day, the conclusive plaque on the Haymarket Memorial Sculpture tells a vague and abstract story:

> *Over the years, the site of the Haymarket bombing has become a powerful symbol for a diverse cross-section of people, ideals and movements. Its significance touches on the issues of free speech, the right of public assembly, organized labor, the fight for the eight hour work day, law enforcement, justice, anarchy and the right of every human being to pursue an equitable and prosperous life. For all, it is a poignant lesson in the rewards and consequences inherent in such human pursuits.*

9/11 AS MODERN SPECTACLE AND MONUMENT

Perhaps the ultimate example of historical spectacle in the context of modernity took place on September 11, 2001. We're still unsure what to call those events: 9/11, The Events of September 11th? We mark the event by its date of occurrence; its place on the timeline. On that date, nineteen men (fifteen from Saudi Arabia, two from the United Arab Emirates, one from Egypt, and one from Lebanon) coordinated attacks on strategically chosen targets in the United States that represented political, military, and economic interests. Affiliated with a politicized Muslim group organized by the oil billionaire Osama Bin Laden, these men hijacked four passenger jets owned by American corporations and flew three of them into both towers of the World Trade Center in New York City and the Pentagon Building in Washington, D.C. The fourth plane crashed in a field in Shanksville, Pennsylvania, though its target was believed to be the White House.

Each of the key elements of modernity, technology, economic liberalization, nationalism, and democratic ideology, resonate in the historical spectacle of 9/11. Technologically, 9/11 could not have occurred without the development of air travel. Humanity has been dreaming of flight since the Greek myth of Icarus, but on 9/11, the hijackers turned the dream into a nightmare, using the commercial aircraft and the fuel they stored as giant missiles. One of the targets of the hijackers, the Pentagon, has come to symbolize the technocratic nature of American military imperialism of the modern era. Also with respect to the technological aspect of 9/11, much of America watched the event occur in real time. Communication technology made the event accessible to anyone with a television. Due to the proliferation of cameras in mobile phones, many eyewitnesses watched the events unfold through their camera lenses, and uploaded their perspectives to the internet. Digitally preserved, we can watch the events unfold from hundreds of points of view.

The market economy was a symbolic target for the hijackers. The World Trade Center Towers were the tallest buildings of the iconic Manhattan skyline, and were symbolic reminders of the dominance of capitalism over the global economy. Designed by architect Minoru Yamasaki, the intentionally modernist towers were the tallest buildings in the world when they were completed in April 1973. The towers represented the overarching power of capital in their design and construction. In planning for the World Trade Center, the New York Port Authority used eminent domain to evict hundreds of tenants previously occupying the real estate upon which the complex was to be built. Upon completion, the architect Lewis Mumford criticized the modernist design as "purposeless giantism" which appeared to be nothing more than "glass and metal filing cabinets."[73] In an economic system that had reduced humans to resources for corporate exploitation, it was a valid criticism.

To the hijackers, the towers symbolized the worst excesses of global capitalism. While many Americans, regardless of their economic status, viewed capitalism as a liberating economic system, they were shocked on that day to find that not everyone agreed. Capitalist pursuits of natural resources in "The Middle East" had wreaked havoc on the political, social, and economic fabric of the region since the end of World War I. That fifteen of the nineteen hijackers on September 11, 2001 came from Saudi Arabia was a surprise to many Americans, as they believed that the relationship between the two nations was mutually beneficial.

The events of 9/11 affirmed a sense of nationalism in the United States not matched since the bombing of Pearl Harbor in 1941. Americans from the Bronx to Boise, Idaho reacted with a sense of nationalist unity, though their circumstances were hardly similar. That nationalism was exemplified by President George W. Bush's assertion before a joint session of Congress that "you're either with us, or you're with the terrorists."[74] The tone turned xenophobic quickly, and "us" was implied as white Christian, while "the terrorists" were implied to be brown and Muslim. Incidents of mosque vandalism and harassment of Arab-Americans increased exponentially. In response to the September 11th attacks, the United States attacked Al-Qaeda, an organization which confounds nationalist identity, first in Afghanistan, and later, controversially, in Iraq. In doing so, the United States showed its dependence on old rules nation-state warfare when the moment called for a more unconventional response. The result was a decade of war that cost upwards of five trillion dollars with little to show in return.[75]

A major frustration shared by many Americans as details of the plot unfolded was that the hijackers had essentially taken advantage of the freedoms of western society in order to attack it. As a result, rights like privacy, freedom of movement, and freedom of religion were called into question in the wake of the 9/11 attacks. President George W. Bush declared in a speech the day after the attacks that "we will not allow this enemy to win the war by changing our way of life or restricting our freedoms."[76] Not long after that, the Bush administration began passing laws that began restricting freedoms on an unprecedented scale, including the Patriot Act of 2001, which allows the government to bypass the fourth amendment; the Military Commissions Act of 2006, which allows for indefinite detention without trial; the FISA amendments of 2008, which allow for warrantless surveillance; and policy changes that allow for ethnic profiling and torture.[77]

Whereas most monuments prompt the viewer's gaze upwards in awe, the 9/11 Memorial at the site of the former World Trade Center in lower Manhattan draws the viewer's gaze downwards, temporarily, into two holes in the ground where the towers originally stood, to commemorate the loss of the structures and those whose lives were lost in them in 2001. Overlooking the Memorial is the newly constructed "Freedom Tower," the plans to which the author and social critic James Howard Kunstler decried as the "eyesore of the month" on his website in July 2005. Kunstler argued that the Freedom Tower's design is "the antithesis of freedom. . . . Instead of a ground floor that offers connection to the pedestrian life of the street, what you get is a gigantic blank-walled crypto-military fortification—two hundred feet of steel and concrete bombproof bombast—while the priapic tower above holds office workers hostage in the world's number one target for shoulder-launched missiles and other weapons of opportunity." Its focus on security rather than freedom is paradoxical, but as a monument serves as an uncannily accurate depiction of the direction in which American considerations of freedom and security have gone since the event that necessitated both the memorial and the new construction. Kunstler concluded that "it's a disgrace to the city of New York and to the word *freedom*."[78]

Official accounts of the 1,776-foot tall Freedom Tower differ sharply with Kunstler's interpretation of it. At the groundbreaking ceremony held on July 4, 2004, New York Mayor Michael Bloomberg declared that "as we lay this cornerstone, we remember that the liberties, which are the bedrock of our nation, can never be shaken by violence or hate." Bloomberg dutifully affirmed the official narrative of the War on Terror, which was at its apex/nadir in 2004: "By laying this magnificent cornerstone of hope, we are sending a message to the people around the world that freedom will always prevail. The war on terror that we now fight requires courage and our freedoms will always be the source of our courage." Site planner David Libeskind asserted that "we commemorate not only the physical restoration of the site, we also celebrate the strength and resilience of the human spirit. This magnificent Freedom Tower ... will inspire New York, America and the entire world with the ideals of liberty and democracy."[79] The memorial is largely nationalist propaganda; no conversation about the event in any global historical context takes place on the site. It makes no mention of American policies in the Middle East since the end of World War I, particularly in Saudi Arabia where the majority of the hijackers were from. Meanwhile, the 9/11 Museum Memorial Store, located in the shadow of St. Paul's Chapel, commercializes the grieving process with a wide array of marketed forms, including t-shirts, coffee mugs, jewelry, mouse pads and cell phone cases for mourning tourists (See Figures 4.8 and 4.9). One can only speculate about the significance of the closest piece of anonymous graffiti declaring that "9/11 was an inside job."

CLEAR HOLD BUILD

Chapter 5, Section 51 of the United States Army Counterinsurgency Manual describes the strategy of "Clear Hold Build" as follows:

> A clear-hold-build operation is executed in a specific, high-priority area experiencing overt insurgent operations. It has the following objectives:
> - Create a secure physical and psychological environment.
> - Establish firm government control of the populace and area.
> - Gain the populace's support.[80]

Figure 4.8 9/11 Memorial Museum Store (exterior view), 2014.
Photograph by Isaac Hand.

Figure 4.9 9/11 Memorial Museum Store (interior view), 2014.
Photograph by Isaac Hand.

The foundations of this military strategy lay primarily at the beginnings of English imperialism in North America. The imperative to clear land was central to English models of political economy since the obliteration of the English Commons at the threshold of the modern era. The goal was to establish a new Olympus, or in the English Puritan John Wintrhop's words: "A city upon a hill."[81] The elements of the Clear Hold Build strategy of appropriating space for imperialist pursuits are evident in early colonial America and in modern America before they become a matter of formalized military policy and exported to Afghanistan and Iraq during the War on Terror.

The native population, already a presence in the contested space, has to be systematically marginalized. Chapter 5, Section 56 of the United States Army Counterinsurgency Manual initially describes the process of Clearing as "a tactical mission task that requires the commander to remove all enemy forces and eliminate organized resistance in an assigned area (FM 3–90). The force does this by destroying, capturing, or forcing the withdrawal of insurgent combatants," and warns that "these offensive operations are only the beginning, not the end state." It goes on to explain that "rooting out such infrastructure is essentially a police action that relies heavily on military and intelligence forces until HN police, courts, and legal processes can assume responsibility for law enforcement within the cleared area."[82] To clear land in the colonies, the English employed both martial and legal means. By trying cases of trespass, theft, and other crimes in which a system of private property is paramount, the English brought Native Americans into an unfamiliar legal system where the natives had little recourse. The historian Francis Jennings explains this process as a "systematic deprivation of property," by which English colonists could take legal title to Indian land by imposing imperial will in courts of law.[83]

If the native population actively resisted the systematic encroachment of English colonists into North America, the gun would often replace the gavel as the primary means of clearing territory for settlement. The myriad battles between Native Americans and English colonists over the course of three centuries, the sixteenth through the eighteenth, had many layers and meanings, but ultimately any agreements drawn up to end them were primarily concerned with respective rights to land (either ceding or respecting territory), and clarifying boundaries. In cases where resolutions were not formalized, victory meant taking territory. If not cleared by gavel or gun, disease would ultimately work its course. Smallpox killed more Native Americans than systematic deprivation or violence, but once the land was cleared, the Christian English took it as a sign of God's providence, and moved in to hold and build on it: to bring lightness to the dark.

Clearing land also has a second, more subtle meaning. Clearing land ordered labor hierarchically, maximized food production in economic terms, and clearly demarcated the "lightness" of civilization from the "darkness" of the forest and their respective inhabitants. The idea of "settling" land was fundamental to the English imperial project. Queen Elizabeth's initial letter of patent to Sir Humphrey Gilbert in 1578 authorized him as an imperial agent "to discover, finde, search out, and view such remote, heathen and barbarous lands, countreys and territories not actually possessed of any Christian prince or people, as to him, his heirs & assignee, and to every or any of them, shall seeme good: and the fame to have, hold, occupie and enjoy . . . to build and fortifie."[84]

Clearing land is primarily an environmental pursuit, one that changes the fundamental dynamic of a particular place to make it conform to the expectations of the imperial directive. In the beginnings of English imperialism in New England and Virginia, this was often clearly demarcated by the fencing of cleared land to pen in animals and keep out Native Americans. Clearing land to make it arable, by felling trees and draining swamps, became a measure of a colony's progress as it pushed outward to expand.[85] Pastoralism and agriculture, the primary forms of economic activity in the nascent colonies, was a major force in the territorial expansion of the early English colonies. That economic pursuit literally changed the landscape of North America.

Chapter 5, Section 60 of the United States Army Counterinsurgency Manual describes the process of holding as follows:

> Ideally HN (Host Nation) forces execute this part of the clear-hold-build approach. Establishment of HN security forces in bases among the population furthers the continued disruption, identification, and elimination of the local insurgent leadership and infrastructure. The success or failure of the effort depends, first, on effectively and continuously securing the populace and, second, on effectively reestablishing a HN government presence at the local level. Measured offensive operations continue against insurgents as opportunities arise, but the main effort is focused on the population.[86]

In early American colonial terms, holding was the pacification of whatever remaining native population was not eliminated by warfare or disease. The most common method of pacification was cultural assimilation through religious indoctrination. By getting the native population to police itself, the English could begin to evaluate the degree to which the natives could effectively contribute to the imperial mission of resource extraction and wealth building. The acceptance of Christian discourse and practice was often rewarded by English imperial agents by bestowing a degree of status or material wealth on compliant natives.[87]

The next step in holding cleared territory, as the United States Army Counterinsurgency Manual states in Chapter 5, Section 61: "key infrastructure must be secured." This is a longstanding tactic of imperial expansion. One of the initial instances of securing infrastructure in the history of British imperialism in North America occurred at the Jamestown settlement in 1607. There, the London Virginia Company's written instructions to settlers insisted that "you shall do your best endeavor to find out a safe port in the entrance of some navigable river," qualifying the directive by advising them to "make election of the strongest, most wholesome and fertile place." To ensure logistical lines with England were not interrupted, the Company insisted that "you must in no case suffer any of the native people of the country to inhabit between you and the sea coast."[88]

After holding the territory that has been cleared, an imperialist agency can begin to reenvision the landscape. Building, as described in the United States Army Counterinsurgency Field Manual, is intended to mean building support for the "host nation" government, assuming that it has developed in accordance with the expectations of American politicians. To do so, Chapter 5, Section 70 advises a wage system to hire local workers to do "beneficial work." It outlines "sample tasks," which include:

- Collecting and clearing trash from the streets.
- Removing or painting over insurgent symbols or colors.
- Building and improving roads.
- Digging wells.
- Preparing and building an indigenous local security force.
- Securing, moving, and distributing supplies.
- Providing guides, sentries, and translators.
- Building and improving schools and similar facilities.

According to Queen Elizabeth's letter of patent to Sir Humphrey Gilbert in 1578, the extraction of wealth was a primary right of the government after having invested in the effort to civilize a place: "any person then being, or that shall romaine within the allegiance of us, our heires and successours, paying unto us for all services, dueties and demaunds, the fift part of all the oare of gold and silver."[89] By having locals build an infrastructure, the imperial architects can begin to orchestrate the process of resource and wealth extraction. The Field Manual offers a case study synopsis of the Clear Hold Build in Tal Afar, Iraq, concluding with a quote from the 3rd Armored Cavalry Regiment Commander, Colonel H. R. McMaster: "The people of Tal Afar understood that this was an operation for them—an operation to bring back security to the city."[90] The Field Manual case study makes no mention of the fact that the Iraq-Turkey oil-export route (also known as the Kirkuk-Ceyhan Oil Pipeline) runs through Tal Afar, and, being one of just two practical export routes out of Iraq, is considered to be one of the most strategically important pieces of oil infrastructure in the Middle East. Until Clear Hold Build was imposed there in 2005, sabotage was creating outages which, according to the Wall Street Journal, were "costing the occupation authority millions of dollars a day in potential revenue."[91] At the time the Clear Hold Build strategy was applied in Tal Afar, KBR, a subsidiary of the American corporation Halliburton with ties to Vice President Dick Cheney, held the exclusive contract for maintenance of the Kirkuk-Ceyhan Oil Pipeline in Iraq.[92]

One of the final steps of Building in the United States Army Counterinsurgency Field Manual is to impose "Population Control Measures." To do so, the military begins by establishing the foundation for a permanent surveillance infrastructure. The manual asserts that "establishing control normally begins with conducting a census and issuing identification cards." The census is designed to establish the framework of social, political, and economic activity in the territory: "Census records provide information regarding real property ownership, relationships, and business associations."[93] The identification cards institutionalize those activities, and allow for their recording and monitoring.

The Counterinsurgency Manual accounts for "other population control measures," as part of the building process, which include:

- Curfews.
- A pass system (for example, one using travel permits or registration cards) administered by security forces or civil authorities.
- Limits on the length of time people can travel.
- Limits on the number of visitors from outside the area combined with a requirement to register them with local security forces or civil authorities.
- Checkpoints.

These measures resonate with the surveillance system imposed on slaves in the English colonies of North America. The roots of the modern surveillance apparatus in slavery has been most ably investigated by Christian Parenti, author of *The Soft Cage: Surveillance in America from Slavery to the War on Terror*. From curfews, slave passes, limits on the length of time people can travel and with whom, checkpoints and slave patrols, all of these are part of a power dynamic that attempts to restrict, control, and monitor mobility, which Parenti calls "the circuitry of resistance."[94] As we have seen in the previous chapter, the idea of circuitry works as a metaphor, but is an increasingly literal part of resistance to technocratic monoculture, with digital surveillance and attempts to circumvent it as prevalent a feature in the twenty-first century as more physical forms were in the nineteenth century.

While Clear Hold Build has worked as an effective strategy for imperialist pursuits in foreign lands for the duration of the modern era, it is equally effective in domestic terms. The process of territorial appropriation known as "gentrification" follows a Clear Hold Build pattern in ways that are undeniably imperialist in form and function. The process of "Clearing" in gentrifying urban landscapes begins with the displacement of the "native" population of a neighborhood: often lower income residents. This is accomplished through the influx of outside wealth, which drives up rents and forces the relocation of a neighborhood's original inhabitants who can no longer afford to live there. Gentrification, a term both ideologically and etymologically rooted in imperialist land policies, has roots in the enclosure system that began in sixteenth-century England, by which common lands were fenced off, *enclosed*, and a system of rents imposed on commoners by landlords who sought to extract wealth from previously uncommodified practices and customs. Clear Hold Build is the apotheosis of imperialist appropriation of space, and gentrification is the form in which it manifests domestically.

TOMPKINS SQUARE PARK

One of the most explicit examples of gentrification in the modern era occurred in 1988 in the East Village of Manhattan. Tensions in the community had risen since the newly minted and ever expansive Yuppies (Young Urban Professionals) of the Reagan Era identified the East Village as an ideal geographical location on the island and began the Clearing process by moving in on the community of artists, homeless people, squatters, and working class families that all lived within a few blocks of Tompkins Square Park, the axis upon which the East Village neighborhood spins. *The New York Times* described the community at the time as "an unconventional lot. . . . Some are survivors of the 1960's anti-war protests, others were dropouts from 1980's materialism," and the neighborhood they called home as a collection of "bombed out buildings and garbage-strewn lots."[95] The media paid a great deal of attention to sensory descriptions of the East Village's inhabitants, referring to them as in articles as a "motley crowd," "a shadowy community," "rowdy drunks and punk rockers with spiked hair and leather garb," purple-hair homeless anarchists" and often focused on their unconventional identities: Jerry the Peddler, Adam Purple, and John the Squatter.[96] These superficial descriptions are not unlike the ways in which early English colonists described the native populations. In sum, there was chaos where Zeus saw a need for order to protect Olympus, so he summoned the Cyclopes from Tartarus.

By economizing the landscape and identifying the need for "improvements" in the economic sense of the term, the colonizers, with their newfound wealth borne of market deregulation and a wholesale assault on the labor unions that had built the postwar middle class, moved in and sought to impose their vision of neighborhood on its inhabitants. Just as the English colonists had centuries earlier, the new presence in Tompkins Square Park sought to exert their power through law enforcement. As a newly invested police force sought to "clean up the streets" of the East Village through selective code enforcement and racial profiling, accusations of arbitrary treatment and police brutality grew. One anonymous local put it quite plainly: "The police are just there to patrol for the real-estate interests. It's the rich against the poor, and the poor are going down the drain."[97] In the meantime, real estate developers were advertising new condominiums with views of the park for upwards of $1 million.[98] Taking control of key infrastructure, according to the Counterinsurgency Manual, is the primary goal of Holding after an initial Clearing has occurred.

As part of the final Building process, a one a.m. curfew had been imposed to keep Tompkins Square Park clear of the inhabitants still standing in the way of gentrification. As a public space, Tompkins Square was previously accessible twenty-four hours a day. On August 7, 1988, 450 riot-equipped police descended on the hundreds of protestors who assembled in the park to protest the curfew bearing banners that read "Gentrification Is Class War" and "Tompkins Square Is Everywhere." The aftermath of the confrontation that followed, which lasted until six a.m., left thirty-eight civilians and seven police officers injured. As a result of the public outcry that followed the next day, Mayor Edward Koch temporarily suspended the police-ordered one a.m. curfew, citing the heat wave that had gripped the city as justification for the reversal. The commander of the 9th Precinct, Captain Gerald McNamara, cited the heat in his passive-voiced response to initial questions about police brutality that came in the aftermath of the riot: "It was a hot night . . . obviously tempers flared. But all these allegations will be investigated."[99]

Things got even hotter in the wake of the Tompkins Square Protests. Allegations of brutality grew and criticism of the police grew. Within two days, twenty-seven charges of police brutality were filed to the Civilian Complaint Review Board, prompting 300 residents to gather at the nearby Saint Brigid's Roman Catholic Church to call for the resignation of Captain McNamara. Those who did not witness the clash between protestors and police firsthand saw what had happened through the proliferation of a relatively newly accessible technology: videotape recording. Citizens had videotaped the clash from various points around Tompkins Square Park, and those videotapes became evidence even Police Commissioner Benjamin Ward could not discount. One commenter called it "the democratization of surveillance."[100]

Not even a week after the smoke had cleared, by which time the number of charges of police brutality had grown to seventy-three, Ward was forced to admit that poor supervision (the highest ranking officer at the scene, Deputy Chief Thomas Darcy, had left to use the precinct bathroom over a mile away when the fighting began) and "questionable tactical decisions" on the part of the police had caused the clash. Those "questionable tactical decisions" included attacking people in organized, planned offensive charges throughout the night with obvious intent to injure, assaulting people who were taking pictures and videotape at the scene, some of whom were clearly wearing Press ID cards, and hurling racial epithets at the protestors and bystanders alike.[101] One editorial writer in *The New York Times* determined it was not the protestors who rioted that night, but the police.[102]

In the aftermath of the violence, Tompkins Square Park continued to be a battleground between indigenous and imperial forces in the East Village. On June 3, 1991, the park was finally cleared of a homeless tent city that had

162 *Of Spectacles and Monuments*

been persistent since the flashpoint of protest three years earlier. This time, the police sweep was "carefully orchestrated," and once again, weather played a critical role. A thunderstorm dispersed a crowd who had begun to protest the eviction, and this time, newly elected Mayor David Dinkins made no apologies for the action: "This park is a park, not a place to live. I will not have it any other way."[103] After the eviction, the city fenced off the park, save for two entrances, for a year. Checkpoints are a critical part of the Building process in Clear Hold Build counterinsurgency. Deputy Mayor Bill Lynch concluded that "this is not what the park is created for. It impedes the use of the park for all the community."[104] By "all the community," Mr. Lynch meant of course the newly invested condominium owners who had paid premium prices to appropriate a neighborhood and replace whatever community had existed with an imposition of wealth, power, and control.

The efforts to gentrify the East Village were largely successful. In 2004, *The New York Times* ran an article in its Real Estate section titled "Bohemia with a Softer Edge: Tompkins Square Park," celebrating the neighborhood's "colorful, countercultural past," while pointing out that "investment bankers and corporate lawyers call Tompkins Square Park home today, but historically, the neighborhood was heavily catered to artists, writers, actors and musicians." At the time the article was written, the median price of a single family apartment in the neighborhood was $750,000, though real estate agent Natalie Pitta insisted that the neighborhood "appeals to iconoclasts."[105]

Nearly one hundred years before the Tompkins Square Riots, the social activist Henry Cogswell commissioned a monument to celebrate the cause of Temperance and had it placed in Tompkins Square Park (See Figure 4.10). The monument is a pyramidal pediment, upon which the Greek goddess Hebe sits. Hebe was the cup-bearer to the gods and goddesses of Mount Olympus. By the time of the riots, the monument was is disrepair, and the statue of Hebe was badly mangled. In 1992, with Tompkins Square Park and the entire East Village cleared, held and built, the monument was restored and a new statue of Hebe was installed.[106] In this continuing Titanomachy, the Olympians had pushed back Chaos once again, and Hebe was returned to serve them.

REFRAMINGS

Spectacles are temporal events that challenge the hierarchical authority of technocratic monoculture with teleological progress. Monuments are spatial objects that exist in official space to co-opt the message of the spectacle and impose a commemoration, which reinforces the very systems that the memorialized events stood to challenge. Here, the ontology of space interrupts teleology. Citizens are rendered subject in monumental space, which severs the connection between the past and present in order to derail the

Figure 4.10 Statue of Hebe in Tompkins Square Park, 2014.
Photograph by Isaac Hand.

prospect of similar future occurrences. The idea of freedom is sacrificed on the altar of security, leaving the observer grateful for the sacrifices of past generations, convinced that continuing the pursuit of such struggles would be both subversive and unnecessary, and that suggesting otherwise would be disrespectful to the past martyrs and the sacrifices they made.

In the meantime, space is physically and ideologically cleared, then rearranged to reflect the interests of modern technocratic monoculture. Once cleared and held through capitalist appropriation and surveillance, the space is remade (built) to reflect the interests and values of the technocratic monoculture, at the expense of all else. The city serves as receptacle for the ideological framework of the nation, and the nation above all advances the concept of security-based dependence, rendering citizens subject once again. It all produces a sort of learned helplessness. As Kate Sopko observes in her 2008 essay *Stewards of the Lost Lands*, "it's no wonder no one wants to think about much. It's hard to think about things that seem completely out of our control."[107] Yet, while things may seem out of our control, they are not. The answer lies in the spaces we live in. To realize teleological time, we must reframe ontological space. The answer is all around us, in the unmappable places that fall outside of the ability of the technocratic monoculture to comprehend, control, and manipulate.

NOTES

1. Hesiod, "Theogony,"*Hesiod and Theogonis*, trans. Dorothea Wender (New York: Penguin Books, 1973), 24.
2. Paul Deman, "Literary History and Literary Modernity," in *Blindness and Insight: Essays in the Rhetoric of Contemporary Criticism* (London: Methuen, 1983), 146.
3. Modris Eksteins, *Rites of Spring: The Great War and the Birth of the Modern Age* (New York: Vintage Press, 1989), 16.
4. Alvin Toffler, *Future Shock* (New York: Random House, 1970), 22.
5. Guy Debord, *The Society of the Spectacle* (Canberra: Treason Press, 2002), 6–9.
6. Ibid. 44.
7. Evgeni Lampert, *The Apocalypse of History* (London: Faber and Faber, 1948), 14.
8. Ibid. 157–159.
9. Guy Debord, *The Society of the Spectacle* (Canberra: Treason Press, 2002), 37.
10. Georg Wilhelm Friedrich Hegel, *The Philosophy of History*, trans. J. Sibree and C. J. Friedrich (New York: Dover, 1956), 442–457.
11. Guy Debord *The Society of the Spectacle* (Canberra: Treason Press, 2002), 45.
12. Ibid. 41.
13. Jacques Derrida, *The Beast and the Sovereign, Vol.1* (Chicago: University of Chicago, 2009), 17.
14. Yi-Fu Tuan, *Space and Place: An Experiential Perspective* (Minneapolis: University of Minnesota Press, 1977), 153–154.

15. Plutarch, *Aristedes*, trans. Bernadotte Perrin (New York: Charles Scribner's Sons, 1901), Chapter 7, Section 2.
16. Ibid. Chapter 7, Section 3–4.
17. Gordon P. Kelly, *A History of Exile in the Roman Republic* (Cambridge: Cambridge University Press), 1–2, 28.
18. Jo Marie Claasen, *Displaced Persons: The Literature of Exile from Cicero to Boethius* (Madison: University of Wisconsin Press, 1999), 252–259.
19. R. B. Dobson, *The Peasants' Revolt of 1381* (London: Macmillan Press, 1970), 159.
20. Ibid. 164.
21. Alastair Dunn, *The Great Rising of 1381: The Peasants' Revolt and England's Failed Revolution* (Stroud: Tempus Press, 2002), 136.
22. Dan Jones, *Summer of Blood: The Peasants' Revolt of 1381* (London: Harper, 2010), 201.
23. R.B. Dobson, *The Peasants' Revolt of 1381* (London: Macmillan Press, 1970), 186.
24. John Milton, "Paradise Lost," in *John Milton: Complete Poems and Major Prose*, ed. Merritt Y. Hughes (New York: Macmillan, 1957), 240.
25. Eric Hobsbawm, *Primitive Rebels, Studies in Archaic Forms of Social Movement in the 19th and 20th Centuries* (Manchester: Manchester University Press, 1959), 13.
26. Stephen Knight, *Robin Hood: A Complete Study of the English Outlaw* (Oxford: Blackwell Publishing, 1994), 218.
27. Marcus Rediker, *Between the Devil and the Deep Blue Sea: Merchant Seamen, Pirates, and the Anglo-American Maritime World, 1700–50* (Cambridge: Cambridge University Press, 1987), 298.
28. Ted Dagne, "Somalia: Conditions and Prospects for Lasting Peace," *CRS Report for Congress* (Washington: Congressional Research Service, August 13, 2011), 15.
29. Michael Wilson, "Dragnet Yields Whimsy and Dread Upstate," *The New York Times*, July 15, 2006.
30. Lou Michel and Dan Herbeck, "Phillips Admits Killing State Trooper Ex-Fugitive Also Pleads Guilty to Wounding Two Other Troopers," *The Buffalo News*, November 20, 2006.
31. "'Guilty as Hell' Pleads Cop Killer 'Bucky' Phillips," *Police One*, November 30, 2006, www.policeone.com/legal/articles/1194278-Guilty-as-hell-pleads-cop-killer-Bucky-Phillips/
32. Danby Pickering, ed., *The Statutes at Large, From Magna Chart to the End of the Eleventh Parliament of Great Britain* (London: Joseph Bentham, 1764) Cap V, Vol. XIII, 142–146
33. Ibid. 146.
34. David J. McCord, ed., *The Statutes at Large of South Carolina. Vol. 7, Containing the Acts Relating to Charleston, Courts, Slaves, and Rivers* (Columbia: A.S. Johnston, 1840), 397.
35. Reinhold Niebuhr, "The Tower of Babel," in *Reinhold Neibuhr: Theology of Public Life*, ed. Larry Rasmussen (Minneapolis: First Fortress, 1991), 84–85.
36. Sanford Levinson, *Written in Stone: Public Monuments in Changing Societies* (Durham: Duke University Press, 1998), 10.
37. Katherine Hodgkin and Susannah Radstone, "Patterning the National Past," in *Contested Pasts: The Politics of Memory*, ed. Katherine Hodgkin and Susannah Radstone (London and New York: Routledge, 2003), 173.
38. Henri LeFebvre, *The Production of Space*, trans. Donald Nicholson-Smith (Malden and Oxford: Blackwell, 1991), 220.

39. "Protest Banners Hung From Acropolis," *Al Jazeera*, December 17, 2008. www.aljazeera.com/news/europe/2008/12/2008121718556743523.html
40. Jerome Roos, "Exarchia and the Greek Spirit of Resistance," *Roarmag*, July 18, 2011, http://roarmag.org/2011/07/exarchia-and-the-greek-spirit-of-resistance/
41. "Bloody Interesting Decision," *Rocky Mountain News*, February 7, 1990, www.coloradoaim.org/history/19900207rmnopedrussellmeanssentence.htm
42. Henri LeFebvre, *The Production of Space*, trans. Donald Nicholson-Smith (Malden and Oxford: Blackwell, 1991), 221.
43. Yi-Fu Tuan, *Space and Place: An Experiential Perspective* (Minneapolis: University of Minnesota, 1972), 37–38.
44. Gil Abousnnouga and David Machin, "War Monuments and the Changing Discourses of Nation and Soldiery," in *Semiotic Landscapes: Language, Image, Space*, ed. Adam Jaworski and Crispin Thurlow (London and New York: Continuum International, 2010), 234.
45. Guy Debord, *The Society of the Spectacle* (Canberra: Treason Press, 2002), 36.
46. Maya Nadkarni, "The Death of Socialism and the Afterlife of its Monuments," in *Contested Pasts: The Politics of Memory*, ed. Katherine Hodgkin and Susannah Radstone (London: Routledge, 2003), 195.
47. Peter Linebaugh and Marcus Rediker, *The Many Headed Hydra: Sailors, Slaves, Commoners and the Hidden History of the Revolutionary Atlantic* (Boston: Beacon Press, 1999), 232–233.
48. Hiller B. Zobel, *The Boston Massacre* (New York: W.W. Norton & Company, 1970), 198.
49. William of Newburgh, *Historia Rerum Anglicarum*, ed. Scott McLetchie (London: Seeley's, 1861), Vol. V, Ch. 20, 653.
50. Robert J. Allison, *The Boston Massacre* (Beverly: Commonwealth Editions, 2006), 63.
51. John R. Ellement "An Upgrade of History," *Boston Globe*, May 11, 2011. www.boston.com/news/local/massachusetts/articles/2011/05/11/upgrade_to_raise_boston_massacre_sites_visual_impact/
52. Robert J. Allison, *The Boston Massacre* (Beverly: Commonwealth Editions, 2006), 64
53. Karsten Fitz, "Commemorating Crispus Attucks: Visual Memory and the Representations of the Boston Massacre, 1770–1857." *Amerikastudien*, Vol. 50, No. 3 (2005): 463.
54. "Memorial Placed On Site Of Tyburn Tree," *San Francisco Call*, Vol. 105, No. 173, May 22, 1909, editorial page.
55. "End of Tyburn," *The Glasgow Herald*, Sept 28, 1964.
56. Alan Brooke and David Brandon, *Tyburn: London's Fatal Tree* (Gloucestershire: Sutton, 2004), 236.
57. "The Tyburn" *Wetherspoon*, accessed June 5, 2014. www.jdwetherspoon.co.uk/home/pubs/the-tyburn
58. Simon Caldwell, "Council Plans New Memorial to Tyburn Martyrs," *The Catholic Harold*, January 30, 2009, http://archive.catholicherald.co.uk/article/30th-january-2009/2/council-plans-new-memorial-to-the-tyburn-martyrs
59. "Welcome to Marble Arch," City of Westminster, accessed October 30, 2013. www.westminster.gov.uk/myparks/parks/marble-arch/
60. John Sugden, "Tecumseh's Travels Revisited." *Indiana Magazine of History*, Vol. 96, No. 2 (June 2000): 164.
61. Logan Esarey, ed., *Messages and Letters of William Henry Harrison* (Indianapolis: Indiana Historical Commission, 1922), William Henry Harrison to Secretary of War, July 5, 1809, Vol. I, 353–354.
62. R. David Edmunds, *Tecumseh: The Quest for Indian Leadership* (New York: Pearson Longman, 2007), 68.

63. Glen Tucker, *Tecumseh: Vision of Glory* (New York: Cosimo, 2005), 293–294.
64. R. David Edmunds, *Tecumseh: The Quest for Indian Leadership* (New York: Pearson Longman, 2007), 205–6.
65. "Haymarket Monument Unveiled," The New York Times, May 21, 1889.
66. "Haymarket Statue Removed," *The New York Times*, July 15, 1900.
67. William J. Adelman, "The True Story behind the Haymarket Police Statue," in *Haymarket Scrapbook*, ed. David Roediger and Franklin Rosemont (Chicago: Charles H. Kerr, 1986), 167–168.
68. "Monument on the Move," *Chicago History*, accessed June 5, 2014. www.chicagohistory.org/dramas/epilogue/toServeAndProtect/monumentOnTheMove.htm
69. John Peter Atgeld, *The Pardon of the Haymarket Prisoners*, June 26, 1893, accessed June 5, 2014, http://law2.umkc.edu/faculty/projects/ftrials/haymarket/pardon.html#REASONS_FOR_PARDONING
70. Paul A Shackel, "Remembering Haymarket and the Control for Public Memory," in *Heritage, Labor and the Working Classes*, ed. Laurijane Smith, Paul Shackel, and Gary Campbell (Oxon: Routledge Press, 2011), 45.
71. Ibid. 46.
72. Tom McNamee, "After 118 Years, Haymarket Memorial to Be Unveiled," *Chicago Sun Times*, September 7, 2004.
73. Alden Whitman, "Mumford Finds City Strangled by Excess of Cars and People," *The New York Times*, March 22, 1967; and Lewis Mumford, *The Pentagon of Power* (New York: Harcourt Brace Jovanovich, 1970), 342.
74. George W. Bush, "Address to a Joint Session of Congress and the American People," *Office of the Press Secretary*, September 20, 2001.
75. Mark Thompson, "The $5 Trillion War On Terror," *Time Magazine*, June 29, 2011.
76. "Bush on Security Efforts," *Washington Post*, Sept 12, 2001, www.washingtonpost.com/wp-srv/nation/transcripts/bushtext_091201.html
77. Brennan Center for Justice, Liberty and National Security Program, "How 9/11 Changed the Law," *Mother Jones*, September 9, 2011.
78. James Howard Kunstler, "Eyesore of the Month: July 2005," accessed June 5, 2014, www.kunstler.com/eyesore_200507.html
79. Phil Hirschkorn, "New York Lays Cornerstone for Freedom Tower," *CNN*, July 6, 2004, http://edition.cnn.com/2004/US/Northeast/07/04/wtc.cornerstone/index.html
80. Field Manual No.3–24: Counterinsurgency (Washington, D.C., Department of the Army: 2006), Chapter 5, 18.
81. John Winthrop, *The Journal of John Winthrop: 1630–1649*, ed. Richard Dunn and Laetitia Yeandle (Boston: Massachusetts Historical Society, 1996), 10.
82. Field Manual No.3–24: Counterinsurgency (Washington, D.C., Department of the Army: 2006), Chapter 5, 19.
83. Francis Jennings, "Virgin Land and Savage People." *American Quarterly*, Vol. 23, No. 4 (Oct. 1971): 540.
84. Rev. Carlos Shatter, *Sir Humfrey Glylberte and His Enterprise of Colonization in America* (Boston: Publications of the Prince Society, 1903), 95–102.
85. William Cronon, *Changes in the Land: Indians, Colonists and the Ecology of New England* (New York: Hill and Wang, 1983), 108.
86. Field Manual No.3–24: Counterinsurgency (Washington, D.C., Department of the Army: 2006), Chapter 5, 19.
87. Daniel Richter, *Facing East from Indian Country* (Cambridge: Harvard University Press, 2003), 143.
88. John Smith, *Travels and Works*, ed. Edward Arber (Edinburgh: John Grant, 1910), Vol.1, xxxiv.

168   Of Spectacles and Monuments

89. Rev. Carlos Shatter, *Sir Humfrey Glylberte and His Enterprise of Colonization in America* (Boston: Publications of the Prince Society, 1903), 95–102.
90. Field Manual No. 3–24: Counterinsurgency (Washington, D.C., Department of the Army: 2006), Chapter 5, 23.
91. Chip Cummins, "Efforts to Secure Oil Pipeline Linking Iraq and Turkey Fail," *Wall Street Journal*, December 13, 2003, http://online.wsj.com/article/SB106867930345823400.html
92. Ruba Husari, "Oil Assessment Work Gets Going in North Iraq," *International Oil Daily*, May 2, 2003, www.iraqoilforum.com/?tag=kirkuk-ceyhan
93. Field Manual No. 3–24: Counterinsurgency (Washington, D.C., Department of the Army: 2006), Chapter 5, 22.
94. Christian Parenti, *The Soft Cage: Surveillance in America From Slavery to the War on Terror* (New York: Basic Books, 2004), 16.
95. Michael Wines, "Class Struggle Erupts Along Avenue B," *The New York Times*, August 10, 1988.
96. Douglas Martin, "Motley Crowd Calls Park Special Place," *The New York Times*, July 9, 1999; Michael Wines, "Class Struggle Erupts Along Avenue B," *The New York Times*, August 10, 1988; Robert McFadden "Park Curfew Protest Erupts into a Battle," *New York Times*, August 8, 1988; and Evelyn Nieves, "Tensions Remain at Closed Tompkins Square Park," *The New York Times*, June 17, 1991.
97. Michael Wines, "Behind the Park Melee, a New Generation Gap," *The New York Times*, August 8, 1988.
98. Michael Wines, "Class Struggle Erupts Along Avenue B," *The New York Times*, August 10, 1988.
99. Robert D. McFadden, "Park Curfew Protests Erupt into a Battle," *The New York Times*, August, 8, 1988.
100. Constance Hays, "Home Videos Turn Lenses on the Police: Police Find Their Actions Tracked by Amateurs With Videocameras," *The New York Times*, August 15, 1988.
101. Todd Purdum, "Melee in Tompkins Sq. Park: Violence and its Provocations," August 14, 1988.
102. "Yes, A Police Riot," *The New York Times*, August 26, 1988.
103. John Kifner, "New York Closes Park to Homeless," *The New York Times*, June 4, 1991.
104. John Kifner, "New York Closes Park to Homeless," *The New York Times*, June 4, 1991.
105. C. J. Hughes, "Bohemia with a Softer Edge: Tompkins Square Park," *The New York Times*, October 31, 2004.
106. Tompkins Square Park Temperance Fountain, *City of New York Parks and Recreation*, accessed June 5, 2014, www.nycgovparks.org/parks/tompkinssquarepark/monuments/1558
107. Kate Sopko, *Stewards of Lost Lands* (Cleveland: The Language Foundry, 2008), 3.

Conclusion
Unmappable Places

TITANOMACHY REDUX

The Titanomachy rages all around us. It is not finished. Those who would resist Zeus's imposition of hierarchical spatial order are Typhon's monsters, but where to strike? The Cyclopes are seemingly panoptic, but they do have blind spots, and there are places within the spectrum of the sociospatial matrix that elude their monocular gaze. These manifest at certain intersections of modernity, and therein lay possibilities for the return of Kronos and the "Golden Age" of commonwealth and mutuality. Marx's call to action in the *Feuerbach Theses* resonates loudly in this context: "The philosophers have only *interpreted* the world, in various ways. The point, however, is to *change* it."[1] The goal of singing in dissonant counterpoint to Hesiod, after all, is one of purification.

Technocratic hegemony maintains and expands its authority through the appropriation of power it exerts by means of spatial triangulation and control. The systematic expansion of geographical knowledge has been a driving force of modernity since Columbus's first transatlantic voyage in the late fifteenth century. The primary means of representing this knowledge, the map, is the elemental means of spatial appropriation. Michel Foucault articulated the omnipresence of power in knowledge. J. Brian Harley considered how maps are a form of power-knowledge in which "the distinctions of class and power are engineered."[2] The result is an authoritative, self-reinforcing discourse regarding the connection between spatial presence, spatial knowledge, and power. Within the power matrix of technocratic modernity, mapping is largely a tool of the public and private interests preoccupied with expanding their degree of control over territory and the people who inhabit it. As Douglas Aberley explained, "in our consumer society, mapping has become an activity primarily reserved for those in power, used to delineate the 'property' of nation-states and multinational companies. The making of maps has become dominated by specialists who wield satellites and other complex machinery."[3] Over the course of modernity, technocrats have drawn maps for the purpose of surveying and extracting resources, and for imposing monoculture.

Increasingly, mapping has moved from the geographical to the ideological, from the material to the abstract. Those interested in maintaining or acquiring power draw detailed maps of opinions, ideas, and behaviors to identify patterns that can then be rationalized, systematized, and manipulated. Whatever can be measured can be mapped, classified, and ordered. Ultimately, the purpose is to render variables into predictable constants, to order chaos in both spatial and temporal terms. As Ronald Abler, John S. Adams, and Peter Gould put it in their 1971 technical manual, *Spatial Organization*, "ordered experiences are those which arouse no questions in our minds," emphasizing that "the need for order must be fulfilled, even if the order must be created where none can be discovered."[4] Mapping eliminates variables by offering contexts and parameters in which a set of conditions can be triangulated, understood and controlled.

Henri Lefebvre's theory of the "production of space" is an imposition which suggests the passivity of the people upon whom this system is imposed. In this construction, space is produced by people who don't have to consume it and consumed by people who don't have a stake in its production. His conclusion is brief; he only abstractly speculates as to "what to do about it," acknowledging the need for a *metaphilosophy* "to uncover the characteristics of a philosophy that used to be, its language and its goals, to demonstrate their limitations and to transcend them." He goes on to point out that "nothing of the old philosophical quest will be abolished in the process." In sum, there will be no conclusion, only *transition*.[5] Within this transitory sociospatial matrix, then, reclamation of agency is necessary for anyone seeking to reframe time and space in useful and productive ways. Space should neither be "produced" nor "consumed." Rather, it should be the result of mutual interest, collaboration, and improvisation.

Despite the exponential advances in technological cartography, which has enabled the mapping of geographical space to exacting degrees, and the proliferation of geographical information systems, which map the phenomena and experiences that occur in that geographical space to be systemized and ordered as well, some things remain unmappable. These are the sources of resistance to technocratic monoculture. They are conceptual rather than perceptual, and as such, remain outside of the cartographic gaze. Lived and experienced from within rather than identified and qualified from without, they are spaces that exist both underneath and above the spatial spectrum that the maps depict. Again with Abler, Adams, and Gould, the only way to map conceptual spaces is to construct ideas about them: "Constructs are ideas about experience which impose preliminary order upon them. . . . Concepts are extremely important because they provide us with a means of manipulating constructs."[6] Conceptual spaces have long avoided such manipulation by virtue of the fact that perceptions of them are often grossly inaccurate, often exposing more about the nature of the beholder than the subject beheld. Conceptions are made jointly and completely by those participating in immediacy of the conditions conceived, whereas perceptions are formed outside of and about the experience being perceived from a distance. Conceptions are definitions, perceptions are opinions.

Here, technocratic preoccupations with accuracy serve to protect the internal concept from external perceptions of it. Perceptions are inherently biased and therefore inaccurate to varying degree. This inaccuracy renders the perceptive model useless to the system it is intended to serve. While J. Brian Harley identified the "silence of maps" with regard to modern power dynamics as a source of injustice toward ignored or marginalized peoples, the silence of maps can also be a source of power for those marginalized groups who have been able to stay outside of the bounds of the cartographic spectrum.[7] In modern technocratic monoculture, triangulation means appropriation, appropriation means control, and control means a loss of autonomy and freedom for those subject to it. Avoiding detection, rather than insisting on recognition, may well be a better long-term strategy for the survival of conceptual spaces in which so much happens in opposition to the technocratic monoculture's attempt "to ensure that *nothing happens.*"

In the 1999 edited volume *Mapping Social Networks, Spatial Data and Hidden Populations*, the anthropologist Merrill Singer employed the rhetoric of imperialism in observing that "from a research perspective, hidden populations are methodologically the opposite of 'captive populations,' such as prison inmates, clinical and hospital patients, students and employees. Hidden populations are generally 'neither well defined nor available for enumeration.' With captive populations, by contrast, the universe is relatively well known . . . Moreover, captive populations are, by comparison, easier to reach."[8] The language, while entirely clinical, informs the nature of technocratic triangulation, which Singer insisted was intended to help people "at risk that may benefit significantly from improved applied social science research."[9] Singer recommended "targeted mapping," using "key informants" and surveillance ("unobtrusive measures") to extract the data necessary to begin to study the hidden population or "target group." This philanthropic imperative is reminiscent of Rudyard Kipling's articulation of "The White Man's Burden," the very ethos of imperialism, and uses some of the same tactics to impose the conditions of captivity onto populations previously hidden from the technocratic field of vision, or as Kipling put it in 1899:

> Take up the White Man's burden-
> Send forth the best ye breed-
> Go bind your sons to exile
> To serve your captives' need;
> To wait in heavy harness,
> On fluttered folk and wild-
> Your new-caught, sullen peoples,
> Half-devil and half-child.[10]

In avoiding such impositions, conceptual spaces exist as borderlands within borderlines. As Peter Jackson suggested in *Maps of Meaning*, a "geography of resistance" emerges from a "plurality of landscapes."[11] Neighborhoods, communities, and regions often defy spatial triangulation; they are difficult to "put down" on maps even though they exist within their bounds. It is the conundrum of the technocrats that Ralph Waldo Emerson observed when he wrote in *Uses of Great Men* that "genius is the naturalist or geographer of the supersensible regions, and draws their map; and, by acquainting us with new fields of activity, cools our affection for the old. These are at once accepted as the reality, of which the world we have conversed with is the show."[12] Protecting these "supersensible regions" from the triangulation and appropriation of "Great Men" then becomes the purpose of those committed to progressive teleological progress. These are often the origins of activism that promote social justice and threaten the hegemonic order of technocratic monoculture in the process. There is a tremendous amount of social, political, and economic latitude within them; put otherwise: *freedom*.

Increasingly, we turn to the "unmappable places" within the sociospatial matrix as a way to hide in plain sight from the panoptic gaze of modernity's technocratic Cyclopes and the monocultural hegemony of the Hekatonkheires. Neighborhoods, communities, and regions share the distinct characteristic of being defined not spatially from the outside, but from within by the people who constitute them and live in them. They are organic, evolving, and overlapping spatial conditions that provide respite from the onslaught of modernity. They are perambulated by the people who at once define them and are defined by them: not surveyed and abstracted by outsiders seeking to impose a hierarchical order onto them. Unmappable places allow those who would resist technocratic monoculture to do so from within its pervasive bounds. As the anonymous authors of *The Coming Insurrection* inveigh, "to cope with the uniformity that surrounds us, our only option is to constantly renovate our own interior world."[13]

NEIGHBORHOOD

Neighborhood is a malleable concept. While the default spatial scale of the neighborhood is hyperlocal, social theorists have applied the discourse of neighborhood to the concept of globalization, and practitioners of Big History refer to "Cosmic Neighborhood" when considering the origins of the universe.[14] Its universal applicability as a concept and a metaphor is largely due to its organic, amorphous character. Lewis Mumford wrote in 1954 that "in a rudimentary form neighborhoods exist, as a fact of nature, whether or not we recognise them or provide for their particular functions."[15] Mumford's observation, in his essay *The Neighborhood and the Neighborhood Unit*, was written to engage the debate regarding whether neighborhoods could be planned "from above," by urban planners, or whether they should occur spontaneously. According to Mumford, the roots of modern urban planning lie in capitalism and technology: "How was it that spontaneous neighborhood grouping, so well defined before the seventeenth century, tended to disappear in systemic new plans . . .? One of the answers to this is the segregation of income groups under capitalism, with a sharp spatial separation of the quarters of the rich and the poor; the other was a technical factor, the increase of wheeled vehicles and the domination of the avenue in planning." Neighborhood life was effectively undermined by these imperatives through the destruction of the dynamic spaces in which they existed, bringing Mumford to the conclusion that "it was easier to lose oneself in the city as a whole than to find oneself in a neighborhood."[16] Ultimately, technocrats rediscovered the idea of neighborhood as an opportunity for the expansion of social work as an official civic undertaking. Mumford argued that political efforts to alleviate social impoverishment and to attempt social integration actually compounded the problems. The result was the hegemonic implementation of "neighborhood units," a concept championed by the sociologist and urban planner Clarence Perry. What followed was a tangle of zoning ordinances and "inconsecutive, irrational 'planning,'" which Mumford concluded required "a more adventurous explanation of alternative solutions."[17]

Alternative solutions to top-down neighborhood planning come in social, economic, and political forms. In order to be an alternative to the conditions Mumford identified, they need to challenge both the technological predisposition of modern space along with the capitalist economic system in which that space is organized. Reclaiming space for pedestrians, lost to automobility and commerce in the twentieth century, is elemental to the revitalization of the neighborhood ethic. As Lewis Mumford put it: "To share the same place is perhaps the most primitive of social bonds, and to be within view of one's neighbors is the simplest form of association. Neighborhoods are composed of people who enter by the very fact of birth or chosen residence into a common life."[18] While pedestrian space is presently the pursuit of every Chamber of Commerce group looking to boost sales for its members (a tacit recognition that the strip malls they once lobbied to zone in the suburbs have failed to deliver the expected return), it is important to recognize that resonant pedestrian space is primarily civic, not commercial. Civic spaces are designed to be communal, not proprietary. Commercial activity is peripheral, not central to participation in and engagement of civic space. Rather, social and cultural activity is central to interaction in civic space. In avoiding commercialization, civic space promotes neighborhood. People in civic space are not consumers, they are citizens. Equal participation in space promotes the democratic ethos that ultimately challenges the inequality modernity promotes. As Jane Jacobs pointed out in *The Death and Life of Great American Cities* (1961), "The self government functions of streets are all humble, but they are indispensible. In spite of much experiment, planned and unplanned, there exists no substitute for lively streets."[19]

All this is not to say that economy is anathema to neighborhood. Rather, the informality and immediacy of a neighborhood economy returns us to premodern forms of exchange. Barter, interest-free borrowing, and other forms of casual exchange defy the principles of capitalism. In more formal economic terms, the resurgence of "buy local" imperatives in places where economies were destroyed by globalization of industrial manufacturing has promoted both a sense of neighborhood and a more dynamic alternative to corporate capitalism upon which those cities once depended. Farmers' Markets, Community Exchanges, and Cooperatives, for example, mix pedestrian space with a proximity to the means of production that Marx argued was crucial for the "simple co-operation" of preindustrial capitalism.[20] Growing interest in worker cooperatives, businesses owned by workers who take part in the decision processes normally reserved for owners and managers, is a form of an economic neighborhood, predicated on the dignity of work and the equality inherent in democratic relations. Janelle Cornwell observes in her 2011 study "Worker Cooperatives and Spaces of Possibility" that workers not only own the means of production in cooperative businesses, they also own the space and time in which their work occurs. This, as she points out, leads to the possibility of "the spatial-temporal organization of *noncapitalist* growth," which challenges the "social construction of space and time through the capitalist mode of production and its characteristic social relations."[21]

At its root, neighborhood is a biblical concept. The word "neighbor" first appears in the Bible in the book of Exodus, Chapter 3, verse 22, and is predicated on the sharing of wealth among the community of Jews who had little in the face of Egyptian oppression. A neighborhood provides a collective identity based on mutuality and commonwealth. The last two commandments deal explicitly with interaction among neighbors. In Exodus, Chapter 20, verse 17, God himself inveighs against the bearing of false witness against neighbors, and asserts that "thou shalt not covet thy neighbor's house, thou shalt not covet thy neighbor's wife, nor his manservant, nor his maidservant, nor his ox, nor his ass, nor any thing that is thy neighbor's." As such, the neighborhood is a place where honesty is the law, and jealousy and envy are outlawed. The eighteenth verse of the nineteenth chapter of Leviticus, the third of five books of the Jewish Torah and the Christian Pentateuch, commands that "thou shalt love thy neighbor as thyself." This is the first of many permutations of "The Golden Rule" or "The Great Commandment." A similar version of this idea appears in Deuteronomy, and also in three of the four Gospels: Matthew, Mark, and Luke. Here, the concept of empathy and compassion are intrinsically bound up with the idea of neighborhood. All of this makes the neighborhood theoretically impervious to the exploitative nature of capitalism. How else does advertising and marketing work, except to explicitly encourage people to break the last commandment?

Covetousness is the root of capital. It confuses the impulses of want and need, so that the two words become interchangeable in the lexicon of consumerism. With the interaction of capital essentially banned within the space of the neighborhood by biblical fiat, where sharing overtakes buying and selling and avariciousness is unacceptable, the neighborhood becomes a theater of resistance against the spatial hegemony of the technocratic monoculture. All of these are ultimately lived relations; they cannot be abstracted and they cannot be mapped. Contrary to the mantra which drives the ethos of corporate capitalism, greed is not good, and it is not compatible with the ideological concept of a local, global, or universal neighborhood.

COMMUNITY

In his 1971 book *Rules for Radicals*, Saul Alinsky identified the community as the primary social space in which to organize people for collective revolutionary resistance against "the system." Here, Alinsky suggested, people could change the world from "within the system."[22] The ultimate purpose for community-based collective resistance against "the system," in Alinsky's words, was to take power from the "Haves" and give it to the "Have Nots." According to Alinsky, "the great American dream that reached out for the stars has been lost to the stripes. We have forgotten where we came from, we don't know where we are, and we fear where we may be going. Afraid, we turn from the glorious adventure of the pursuit of happiness to a pursuit of an illusionary security in an ordered, stratified, striped society."[23] In sum, Alinsky sought to reclaim hierarchically ordered, ontological space for the teleological pursuit of liberty. To do so, he sought the social space of community to challenge the political and economic space of autocracy and corporate capitalism.

It is no wonder then, when Barack Obama began campaigning for president in 2007, conservatively minded commentators were quick to simultaneously criticize and dismiss his early work in Chicago as a community organizer. Byron York's article in *The National Review*, titled "What Did Obama Do As A Community Organizer?" complained that Obama encouraged people in poor income situations to act collectively in their self-interest, find issues that the community seemed to agree were worth addressing, and sought a degree of redress through a collective assertion of political power. York concluded in his article that Obama actually didn't accomplish much as a community organizer, and as such, "we shouldn't expect much to actually get done."[24] Sean Hannity addressed Obama's role as a community organizer on his Fox Television show in a special segment called "Obama and Friends: History of Radicalism." In it, his guest Andy Martin, who Hannity described as "a Chicago based internet journalist," asserted that "a community organizer, in Barack Obama's case, was somebody that had been training for a radical overthrow of the government." At the same time, Martin dismissed Obama's time as a community organizer in Chicago, saying that "he had virtually no impact in his so-called community organizing career."25 Again, community organizing was simultaneously a perilous threat to the nation and a failed idea.

With such bipolar framings, it becomes clear that the agents of technocratic monoculture do not have the ideological, conceptual, or rhetorical tools to deal with the concept of community. Like neighborhoods and regions, they are not defined from the outside, but from within. They are not imposed from the top down, but composed from below. The concept of community flusters the Cyclopes, because it lies outside of their range of vision. The Invisible Committee begins the chapter entitled "Get Organized" in *The Coming Insurrection* with the argument "that individuals are possessed of so little life that they have to *earn* a living, to sell their time in exchange for a modicum of social existence." The Committee offers "the exigency of the commune . . . to free up the most time for the most people" as an antidote, suggesting that "there will be no more time to fill, but a liberation of energy that no 'time' contains."26

Commons, Commune, Communism, Community. These words all share the same root, *com-*, a Latin variant meaning "together" and "completely." Their differences suggest the evolution of social, political, and economic organization over the course of modern history. Commons, in the medieval sense of the term, refers to a particular space upon which a group of people negotiated collective access to the hydrocarbon resources necessary for survival. Commons were generally autonomous and self-governed cooperatives. While their existence was institutionalized and protected by law, beginning with the Magna Carta and the Charter of the Forest in 1215 and 1217 respectively, the disappearance of the commons through the process of enclosures, which signaled the beginning of this modern era, led to a politicized attempt at reclamation: communism.27 As with any—*ism*, communism's inherent flaws were rooted in the hierarchical power structures that sought to implement it as official state policy. Politics promote *-isms*, a suffix that denotes systemic doctrine, and justifies the expansion of official authority in the process. Communism, capitalism, etc., are always problematic because of their dogmatic rigidity and their need to perpetuate problems within the system to justify their very existence. Politicians and administrators don't solve problems, they create them in order to maintain and expand their authoritative reach. To borrow from the science fiction writer Jerry Pournelle: "In any bureaucracy, the people devoted to the benefit of the bureaucracy itself always get in control, and those dedicated to the goals the bureaucracy is supposed to accomplish have less and less influence, and sometimes are eliminated entirely."28

Where politicized communism failed, social community triumphs. As the sociologist Ferdinand Tonnies argued in his 1887 work, *Gemeinschaft und Gessellschaft (Community and Civil Society)*: "Thus all community-type law is to be understood as a creation of the reflective human spirit."29 Natural law, rather than the "artifice" or civil law, drives the existence of community. Civil law is codified, natural law is improvised. The malleability of community accounts for its persistence over time and space. It is often in the crises that result from political incompetence or economic trauma that community manifests most clearly. For example, historian Robert McElvaine has observed a revival of "the values of community, justice and cooperation" during the Great Depression.[30] In the electrical blackout of 2003, journalists chronicled the sense of community that was readily apparent in New York City. From an interview in *The New York Times*: "It was amazing last night, the scene. There was a sense of community and excitement in the streets. People were free of technology, having parties with their neighbors or lying around in the grass."[31] Here was community and commons, free of the trappings of modernity. Here was vacation and recreation: a vacating of modern capitalist responsibility and obligation; a recreation of the premodern conditions that Hesiod associated with Kronos and the Golden Age.

Debates about the new nature of community with the proliferation of the internet add to this malleability. Virtuality does not destroy community, though it certainly complicates it. Shared space within the context of modernity has always been a relative concept; observations of transatlantic or international communities imply a certain consciousness of space. The conditions of diaspora and deracination have expanded the landscape of community; virtual space is an extension of this. More than anything, technology enables the proliferation of a community based on ideological rather than geographical terms. Robert Putnam's lament for the decline of community in his 2000 work, *Bowling Alone: The Collapse and Revival of American Community*, rightly called into question the isolating effects of technology on traditional notions of community as shared interaction in common space, but it is simultaneously impossible to ignore the effect of technology in organizing the wave of political protests in the nascent digital age, from Seattle in 1999, to the Iranian Green Revolution in 2009, Tehrir Square and the Arab Spring in 2010, and the Occupy Movement of 2011. Technology cannot replace community in physical space, but it can enhance it and make it a more dynamic force for progressive teleological change.

REGION

In his 1998 volume *Rethinking the Region: Spaces of Neoliberalism*, John Allen observed that regions are "conceived as a series of open, discontinuous spaces constituted by social relationships which stretch across them in a variety of ways."[32] Where are the dividing lines that separate one neighborhood from another? Where does one community begin and another end? Who can draw an accurate, continuous map of the Rust Belt, or the Bible Belt, much less the Middle East or the Midwest? Such things are impossible because of the complexity of human interaction that takes place within them. They are constantly changing in dynamic and unpredictable ways, which makes spatial triangulation from the outside impossible. The supersensible regions that Emerson claimed would make a genius of any naturalist or geographer who could draw their map are impossible to define in temporal and spatial terms, and offer a unique opportunity to resist technocratic monoculture from within. Borderlands, while vital to the formative resistance of nascent modernity and its attendant systems, are no longer essential in the physical sense. They exist conterminously within the sociospatial matrix of precisely defined spaces and places. They are as omnipresent as the panoptic gaze to which they are subject, yet fall outside the visual spectrum of the systems that support it.

The origins of the region as a concept are inextricably linked to the origins of the word. The root of the word, *reg-* is a Latinate variation of *rex*, meaning "king." The Latin word *regio* has broad meaning, and was used to describe a realm, portion of a country, or a territory, or even a climate. Like the dynastic realm that Michael Biggs described as the preeminent form of political spatial organization in medieval Europe, the meaning of region is vague and in flux. Regions, by definition then, are at their root political, but also have cultural, economic, and environmental implications. Surely, regional culture is something that can be oversimplified, but the environmental and economic conditions of a region undoubtedly contribute to the conditions of their respective social landscapes. Like neighborhoods and communities, regions are very difficult to define from an external perspective. As a result of this complexity, J. Nicholas Entrikin observed in his 1989 essay, *Place, Region and Modernity*, that "the study of regions has moved toward the periphery of social science and beyond" due to the fact that "the modern model of scientific rationality has been drawn from the physical sciences, and regional studies have not conformed with this model."[33]

Because a region cannot be rationalized, ordered, and bound by precisely determined borderlines, it remains outside of the technocratic gaze, and thus it is an opportune space for a geography of resistance to manifest. In regions, the cultural, political, economic, and environmental conditions are all intertwined and circular. The culture of a particular region contributes to its political character, while the environmental conditions help define the economics of a region, which in turn define that culture, and the cycle perpetuates itself accordingly. There is a culture of politics and a political culture. At the same time, we can consider the economization of politics and the political economy. Economic culture and the culture of economics are issues that generate thoughtful reflection. In such spaces, Entrikin observed that the boundaries between objectivity and subjectivity, rationality and irrationality, and "the demarcation between scientific knowledge and other forms of knowledge are blurred."[34] Regional identities reflect what the anthropologist Clifford Geertz called "primordial attachments," which tend to complicate officially defined civic identities.[35]

Where science is blind, art is visionary. The Regionalist Art Movement of the 1930s, born in the wake of world war and economic crisis, was an expressive reaction to modernity and the alienation it wrought. Charles

Burchfield, whose work is defined by the unmappable place of regionality and the unmappable time of seasonality, came to comprehend a certain "poetry of place" over the course of his life. Though Burchfield considered himself "not an Ohio, or a western New York artist, but an American artist," his work was, above all else, a meditation on place.[36] At once industrial, residential, and natural, Burchfield's painting reflected a quality that was inherently "local." Simultaneously real and surreal, it was the product of peripatetic rather than catalogued knowledge: time spent in place. While it may be difficult to draw an authoritative map of a region from the outside looking in, anybody can tell you what region they come from, and what makes it unique. It is an identity that is defined by the people who live there, not by boundary lines drawn around it from the outside.

Attempts to triangulate regions from the outside are based on hierarchically authoritative models with inherently imperialist designs. Like neighborhoods triangulated by police officers, and communities monitored by Department of Homeland Security agents, regions as organic entities are increasingly threatened by technocrats acting on the behalf of capitalists. Appointed officers in invented "Offices of Regional Development" who seek to expand the authority of the state as a means to impose capitalist market economics are increasingly present on the bureaucratic landscape. These attempts at globalizing the local and localizing the global make regions the spatial heat shield of resistance to technocratic monoculture. For example, The U.S. Economic Development Administration (EDA), has a program called "Know Your Region," which offers a curriculum promising to explore "regional and local approaches to economic innovation and competitiveness across the United States . . . intended to help local officials, economic development practitioners, community leaders and citizens assess local and regional assets, needs and visions in a global context, leading to long-term regional prosperity and sustainability."[37] Technocratic monoculture is trying to triangulate and commodify the region.

Because of the disparate nature of regional identity, the region is perhaps the most vulnerable to the triangulatory pursuits of the technocratic monoculture. Because it is the biggest of the three "unmappable places," regions represent an opening that, once figured out, could be applied to render communities and neighborhoods subject to the same types of authority as other official models of ontological spatial control. Determining a way to defend the "primordial" nature of the region as an organic space defined by the people who live in it rather than from the outside then becomes a matter of paramount importance. The answer lies largely in economic pursuits. Participating in local economies, rather than encouraging and accepting outside investment through politically organized tax-loophole schemes and incentives, offers regions an opportunity to exercise not only a degree of economic independence, but cultural independence as well. Local foodways, local artists, local businesses, and local politics are all crucial and vital to the cultural definition of a region.

180 *Conclusion*

Localizing the economy issues a larger challenge to the environmental implications of global modernity. James Howard Kunstler questions the sustainability of a global consumer culture, wryly characterized by his idea of "the 3,000 mile Caesar Salad" in his essay *Clusterfuck Nation: A Glimpse into the Future*.[38] By Kunstler's estimation, there is no sustainability in the globalized capitalist economic model, due to its dependence on fossil fuels, which are running out and have no viable replacement to continue the scope and scale of technological mobility as we currently depend on it. Kunstler suggests that regions will move from the abstract to the real very quickly, dependent on their environmental conditions. Access to fresh water, weather patterns suitable for agriculture, and a purposefully designed infrastructure that accommodates rather than obliterates the natural landscape will all bring regional conditions to supersede the state and national political boundaries under which they currently exist. To those who depend on the conveniences of modern life for their economic, political, and social wellbeing, Kunstler's vision of a society felled down to earth like a modern day Icarus invokes visions of a "Dark Age," but to those who seek liberation from the inherent artificiality of global modernity, it represents an opportunity to live closer to the earth, closer to our neighbors and closer to the existential truths that our present age obfuscates in the name of political and economic identity.

CHANGE FROM WITHIN OR WITHOUT?

"Revolution" and "Evolution" are two words that generate a great deal of ideological fervor. Both are necessary factors in any consideration that seeks to result in a measure of change. Which is more effective, which is more practical? Which is best applied to the social, political, and economic conditions that are the target of such charged discourse? Both words are rooted in the Latin *volvo*, which means "to roll." To evolve means to roll forward, to revolve means to roll around. While both imply a degree of change from a present condition, the former is linear and directional, the latter is circular and cyclical. While revolution is on the tongues of the politically disaffected around the world, it promises to return things to a state that arguably never was; a false pretense of past tense idealism that only frustrates participants when it manifests the very conditions it sought to replace. The history of modern revolution is the history of retrenchment. The American Revolution produced a plutocracy of merchants, the French Revolution produced Napoleon, the Russian Revolution produced Lenin and Stalin, and so on, all the way down to the Egyptian Revolution producing the same type of martial government that existed before the protests in Tehrir Square.

In May 1905, the English intellectual Thomas Alfred Jackson wrote an essay in *The Socialist Standard* titled "Evolution by Revolution," in which he concluded that "when the workers shall have evolved so far that they conquer for themselves the political power, that day will mark the realisation of the Social Revolution."[39] Ultimately, Jackson's vision of both phenomena (evolution and revolution) took place within the contexts of the prevailing political, economic, and cultural conditions of the time. Consider the ideological implications of simply switching the words of the title to read: "Revolution by Evolution." Here, an entirely different construct emerges. Rather than assuming the mantle of state and industrial power, "the workers" transcend them altogether. In building new relationships, new means of political, economic and cultural interaction, evolution produces new forms that, to borrow again from Emerson, "cools our affection for the old."

An understanding and engagement of unmappable places becomes critical for such an evolutionary model of change to be possible. To transcend the state, a sense of neighborhood must replace the official apparatus of violence that enforces law. To transcend capitalism, the mutuality and commonwealth of community must replace the principles of appropriation and exploitation driving the consumer economy. To transcend the exclusive and provincial cultural discourse of nationalism, an understanding of regionalism as a continuously porous and fluid borderland of people and ideas must manifest. Such models allow us to work outside of the context of modernity: both above and below, in its foreground and its background, everywhere external to the project. Remember Lefebvre's idea of transition rather than conclusion. These unmappable places exist within the sociospatial matrix of modernity, forming the foundations of an almost parallel reality. In doing so, their conditions go largely unmonitored and hence unrecognized by the agents of technocratic monoculture. Gil Scott Heron's assertion in 1970 that "The Revolution Will Not Be Televised" was especially prescient. The Cyclopes' range of vision is defined by modernity; it operates solely within its parameters and boundaries. Everything external to that range of vision as a means of teleological progress, therefore, is both practical and possible.

This is not to say it is practical or desirable to live entirely "outside of the system." Even John Zerzan, the anarcho-primitivist intellectual, has a website, and once talked about his essay on the importance of silence during his "Anarchy Radio" program. Similarly, the bicentennial conference commemorating the uprising of the Luddites was most heavily advertised on the internet. The triangulations of modernity are so complete that we often must also be "of" it by practical design, even while simultaneously seeking to resist it. Nowhere is really "off the map." As such, teleological existence within the confines of modernity should seek to undermine its more pernicious features. Technology should promote human interaction, not isolation. Market economics should promote the distribution of wealth, not its consolidation. Nationalism should promote cultural understanding, not enforce cultural hegemony. Bureaucratic interactions should be humanized. Social justice should be a priority, not a coincidence. Our choices of consumption should be primarily informed by a concern for the welfare of everyone and everything involved in the means of production and the environment that provided the raw materials of the eventual "product" in question.

Monitoring the monitors becomes another obligation for evolutionary reform. Consider Umberto Eco's assertion in his 1967 essay "Reports from the Global Village":

> So for the strategic solution it will be necessary, tomorrow, to employ a guerilla solution. What must be occupied, in every part of the world, is the first chair in front of every TV set (and naturally the chair of the group leader in front of every movie screen, every transistor, every page of newspaper). If you want a less paradoxical formulation, I will put it like this: The battle for the survival of man as a responsible being in the Communications Era is not to be won where the communication originates, but where it arrives. Precisely when the communication systems envisage a single industrialized source and a single message that will reach an audience scattered all over the world, we should be capable of imagining systems of complementary communication that allow us to reach every individual human group, every individual member of the human audience, to discuss the arriving message in the light of the codes at the destination, comparing them with the codes at the source."[40]

Here, to elaborate on Derrida's construction of the core and the periphery, the beast watches for the incursions of the sovereign into its territory. In doing so, the beast simultaneously subverts and inverts the traditional hierarchical paradigm of power relations. Technocratic monoculture is subjected to observation by its constituency. Critical analysis weakens its power, rendering it marginal and superfluous. It becomes the object and subject of jest; something reified power has never quite come to terms with. From the clever mockery of medieval court jesters, to the biting wit of Gilded Age satirists, to the raw criticism of millennial stand-up and sketch comedians, the comic frame reveals the official as pompous and the wealthy as detached from reality. Mikhail Bakhtin wrote of the concept of *carnivalesque*, and finds the example in the medieval Feast of Fools, held as an anticelebration of the Feast of the Circumcision on first day of the year. During the Feast of Fools, humor, parody, and irreverence were used to challenge the legitimacy of religious, political, and economic authority. This type of folk humor was the manifestation of what all the celebrants knew to be true during the rest of the year: that the religious authorities employed tactics of fear and oppression to maintain their status as the arbiters of morality and order.[41] It is the stuff of comedy; one of the few discursive venues for pure truth to come to light. When the sovereign wanders into the forest of the beast claiming authority, his slippers and robes are muddied and his crown is rendered crooked. His vulnerability in the "real world" over which he claims dominion is nothing if not entirely laughable.

Such informed tactics of intrusion, accommodation, and resistance come from education, which is at present a contested concept. Should education challenge the conditions of technocratic monoculture, or should it reinforce them? Should it make people aware of the world they live in, or simply aware of how they fit into its economic functioning? Do educational institutions exist to produce global citizens or prospective corporate employees? The former conditions rely on education that encourages civic activity, while the latter promotes training to encourage economic activity. Education promotes political, economic, and social evolution, or put otherwise, teleological change. Training ensures an ontological stasis, and a pattern of repression, frustration, and contempt that promotes a violent revolutionary ideology, which ultimately reinforces the very systems it seeks to undo. Talk of revolution encourages the expansion of the "repressive state apparatus." *Hekatonkheiromachy akolouthei.*

Within the "repressive state apparatus," and the technocratic monoculture it supports, the space exists to affect change. Strangely, the systems that depend on ontological stasis for survival use the language of teleological evolution as window dressing for the maintenance of the status quo. Just as many of the signers of the Declaration of Independence had no real belief in the phrase "All men [or better put: *'All People'*] are created equal," the phrase nevertheless provided an opening for the civil rights movement to pursue equality to varying degrees of success over the course of American history. Similarly, the technocratic monoculture's fascination with the rhetoric of evolution provides an opening for agents of progressive teleological change to act within its bounds, openly and without apology. An internal memo, a coordinated media attack, a boycott, or a shareholder's revolt can do more to change the culture of a corporation than a bomb threat or a smashed window ever could. Violence legitimizes a repressive response. Nonviolent but effective action demands accommodation.

These "unmappable places," both internal and external to the technocratic monoculture, provide safe haven for collective resistance. Revolution by force is a romantic, but also a losing proposition; the state is armed to the teeth and has monopolized the morality of violence. Revolution by Evolution, to borrow from Thomas Alfred Jackson, seems a more practical and effective means of challenging the technocratic monoculture on a holistic and systemic basis. Affecting change from within the system is one way of undoing the feedback loop of the present. Simply opting out of the pervasiveness of modernity by choice, when the opportunities present themselves, is an equally effective, if less romantic, means of achieving the levels of freedom and independence more in step with teleological progress than eschatological repression. It is a subject that Albert Camus considered in *The Rebel: An Essay on Man in Revolt*, in which he differentiated between the conditions of revolution and rebellion. Rebellion, as opposed to revolution, is a moderate discourse. "This law of moderation equally well extends to all the contradictions of rebellious thought. The irrational imposes limits on the rational which, in its turn, gives way to moderation." According to Camus, "this limit was symbolized by Nemesis, the goddess of moderation and the implacable enemy of the immoderate."[42] In Hesiod's *Theogony*, Nemesis is also the goddess of Indignation (the resentment of disagreeable conduct), borne of Night, sister of the Destinies, the Fates, and the tribe of Dreams.[43]

This is not to say it is easy to resist. Resistance involves a deliberate process of unlearning. It is a comprehensive reenvisioning of our landscape, our resources, and ourselves. It is the unlearning of consumerism and the learning of citizenship; the unlearning of self-interest and the learning of collective interest; the unlearning of "mine" and the learning of "ours." It is not impossible, but it is difficult. There are degrees of effort involved.

APPROPRIATING MODERNITY

Reshaping the temporal condition of modernity is inherently linked to reimagining the space in which it occurs. It lies on the principle of democratic ideology to appropriate the other elements of modernity—technology, the dual forces of urbanization and nationalism, and the expansion of the market economy—to achieve this. By democratizing these, the teleological cycle that moves us forward while at the same time returning us to the ideals of commonwealth and mutuality that Hesiod articulated in the Titanomachy becomes manifest. The rejection of technocratic monoculture, along with the apocalypticism embedded in its ideological structure, then becomes a critical means of resistance.

The mitigation of technology is essential to the process of rupturing technocratic monoculture. Challenging automobility, for example, results in new kinds of public space, which in turn effect perceptions of time. Limiting remote digital interaction with our jobs returns a physical and ideological sense of places of employment, making time away from the workplace more meaningful. Reimagining the value of technology is as important as rethinking our dependence on it. As new technology is developed, old forms are often discarded as worthless and obsolete. Expropriating old technologies, from printing presses to industrial manufacturing equipment, and repurposing them reconfigures value not in newness, but in endurance. Staying power, not transitory consumption, repositions value in a way that technocracy cannot account for. This is presently occurring on an impressive scale in the Rust Belt of the Great Lakes region in the United States, where it is beginning to become economically possible to build a durable, dynamic community around the cast-off remnants of industrial capitalism.

Nationalism is a difficult thing to challenge; its ominous pervasiveness as the principle means of organizing global society is overwhelming, and nations retain the monopoly on violence. However, the Cyclopes' walls, borderlines imposed on old borderlands, are built on shifting mudsills. Benedict Anderson has shown us that nations and states are imagined, but what is real? Neighborhood, both local and global is real. Community is dynamic and difficult to see from the outside. Regions are interstate and transnational. Both are virtually unmappable, and fall outside of the Cyclopes' visual spectrum. These must be the principle spaces of the temporal resistance. They are what Yi-Fu Tuan characterized as "real places," centers of meaning and experience.[44] They are defined by the people who perambulate them, not by abstractors; they are autonomous and self-organizing, not bureaucratic and imposed. Similarly, urbanization is a defining aspect of modernity. Here, a shift from a consumer culture to producer culture becomes the key to breaking the cycle of dependence upon which technocratic modernity operates. "DIT" (Do It Together) movements in the spatial remnants of cities capitalism has abandoned are using a sense of neighborhood and community to renovate space toward regional productivity. Urban gardens, artisanal workshops, and sustainable development models all create a source of civic pride that a global investment scheme could never replicate. To borrow again from Emerson, "these are at once accepted as the reality, of which the world we have conversed with is the show."

Inside of the increasingly pervasive market economy, cooperative models of economic exchange signal Kronos's steady teleological advance, and offer an option to the unsustainable forms of capitalism advanced by technocratic monoculture. The cooperative model differs from the collective in that it acknowledges the market economy and proprietary ownership, but subverts them from within by disavowing profit and making ownership a function of labor.[45] In a cooperative model, the ethical obligations to everyone involved in the transaction, both the producer and the consumer, are paramount. The environmental spaces in which the entirety of the transaction occurs, from the procurement of raw materials, to the exchange itself, to the manner in which the product is ultimately discarded, reused, or recycled are taken into account. By comprehensively remodeling the spaces in which the market economy occurs (with consideration not only for the means of production, but also the means of consumption and disposal) more equitable, just and sustainable conditions result.

Ultimately, any realization of these conditions requires a critical mass to push back against the prevailing forms of hierarchically framed interaction and discourse in which they manifest. Since the near turn of the millennium, we have seen Kronos's teleology rearing itself with renewed persistence, from Seattle, to Paris and Greece, to Tehran and Tunisia, to Tehrir Square and Tompkins Square, to Detroit and Datong, to Madison, Chicago, and Columbus. If these manifestations of teleological emancipation from centralized, hierarchical systems are to be effective, and not dissolve into the power grabs that have plagued nearly every revolution of the modern era, new models of social, political, and economic interaction must take hold. They must be based on cooperative, not competitive, principles. These will lead to a fundamental reshaping of our understandings of space, which will in turn lead to reclamation of time from stasis toward progress: from eschatology toward teleology.

186 Conclusion

Having begun with Hesiod's written account of the Titanomachy, it should be noted that prior to his transcription in the seventh century BC, the story would have been recounted by a storyteller around a fire, accompanied by ritualized song and dance. Hesiod notes that the Muses sang "unearthly music" acknowledging "the august race of first born gods," of which Kronos would have been among. In that spirit, from one mythological space to another, we conclude in 1972 at the Golden Torch dance hall in Stoke-on-Trent, England. Working class youth, facing bleak economic prospects as the potteries replaced workers with machines, turned their ears westward to the music of African American youth, full of the same restless energy borne of equal parts frustration and hope. Transatlantic neighborhood, community, and regionalism were the foundations of the Northern Soul Movement, which challenged hierarchical notions of race, class, and gender on the unmappable space of the dancefloor, in the unmappable confluences of the scene and the myriad communities that comprised it. It was there that the kids defied conventional notions of time by dancing all night, many fueled by the amphetamine drugs meant to keep them alert at the workplaces that no longer offered much opportunity for employment.

One of the songs they would have heard was "Time" by Edwin Starr, which captures perfectly the frustrations of technocratic modernity as an interruption of teleological progress. The pace of the song is frenetic, 133 beats per minute, roughly the same as the target heart rate for an optimal workout, while the steady bass line punctuated the equal parts resilience and urgency of Starr's fiercely teleological message. After observing the totality of the ontological stasis that had left so many frustrated and alienated from modern society at the end of the 1960s, Starr delivered the ancient promise: "The day when time brings on that precious change, on that day the whole world will sing, time will have washed all our fears away, and peace will have brought us all a brand new day."[46] It is a message of hope that resonates for those on the side of Kronos. *Titanomachy tha teleiósei.*

NOTES

1. Karl Marx, "Thesis on Feuerbach," in *The Viking Portable Marx*, ed. Eugene Kamenka (New York: Viking Penguin, 1983), 155–158.
2. J. Brian Harley, "Deconstructing the Map." *Cartographica*, Vol. 26, No. 2 (Spring 1989): 7.
3. Douglas Aberley, *Boundaries of Home: Mapping for Local Empowerment* (New York: New Society Publishers, 1992), 1–7.
4. Ronald Abler, John S. Adams, and Peter Gould, *Spatial Organization* (Englewood Cliffs: Prentice Hall, 1971), 8.
5. Henri LeFebvre, *The Production of Space*, trans. Donald Nicholson-Smith (Malden & Oxford: Blackwell Publishing, 1991), 406–408.
6. Ronald Abler, John S. Adams, and Peter Gould, *Spatial Organization* (Englewood Cliffs: Prentice Hall, 1971), 13–14.

7. J. Brian Harley "Deconstructing the Map, *Cartographica*, Vol. 26, No. 2 (Spring 1989): 14.
8. Jean J. Schensul, Margaret D. LeCompte, Robert. T. Trotter II, Ellen K. Cromley, and Merrill Singer, *Mapping Social Networks, Spatial Data and Hidden Population* (Walnut Creek: Altamira Press, 1999), 129.
9. Ibid. 134.
10. Rudyard Kipling, "The White Man's Burden," in *Kipling Stories and Poems Every Child Should Know*, ed. W. T. Chapin (Cambridge: Riverside Press, 1909), 359.
11. Peter Jackson, *Maps of Meaning* (London: Taylor & Francis, 1989), 153, 171.
12. Ralph Waldo Emerson, "Uses of Great Men," in *Representative Men: Seven Lectures* (Boston: Houghton Mifflin and Company, 1888), 21.
13. The Invisible Committee, *The Coming Insurrection* (Los Angeles: Semiotext(e): 2009), 39.
14. Fred Spier, *Big History and the Future of Humanity* (Malden: Wiley-Blackwell, 2011), 65.
15. Lewis Mumford, "The Neighborhood and Neighborhood Unit." *The Town Planning Review*, Vol. 24, No. 4 (January 1954): 257.
16. Ibid. 258–259.
17. Ibid. 269.
18. Ibid. 257.
19. Jane Jacobs, *The Death and Life of Great American Cities* (New York: Modern Library, 1961), 156.
20. Karl Marx, *Capital: A Critical Analysis of Capitalist Production* (London: Swan & Sonnenschein & Co., 1902), 326
21. Janelle Cornwell, "Worker Cooperatives and Spaces of Possibility: An Investigation of Subject Space at Collective Copies." *Antipode*, Vol. 44, No. 3 (2012): 726–727.
22. Saul Alinksy, *Rules for Radicals* (New York: Vintage Books, 1971), xix.
23. Ibid. 196.
24. Byron York, "What Did Obama Do As a Community Organizer?" *The National Review*, Sept. 8, 2008
25. Sean Hannity, "Barack Obama, Community Organizing Years," April 29, 2011. *Hannity's America Video Archive*, Fox News Corporation, www.foxnews.com/hannitysamerica/.
26. The Invisible Committee, *The Coming Insurrection* (Los Angeles: Semiotext(e), 2009), 69.
27. Peter Linebaugh: *The Magna Carta Manifesto: Liberty and Commons for All* (Berkeley: University of California, 2008), 142.
28. Jerry Pournelle, "The View from Chaos Manor," *View 406*, April 3–9, 2006, www.jerrypournelle.com/archives2/archives2view/view408.html#Iron
29. Ferdinand Tonnies, *Community and Civil Society*, ed. J. Harris (Cambridge: Cambridge University Press, 2001), 211.
30. Robert McElvaine, *The Great Depression: America 1929–1941* (New York: Crown Publishing, 2009), 205.
31. Susan Saulny "The Blackout: Sense of Community," *New York Times*, August 16, 2003.
32. John Allen, *Rethinking the Region: Spaces of Neoliberalism* (New York: Routledge, 1998), 3.
33. J. Nicholas Entrikin, "Place, Region and Modernity," in *The Power of Place: Bringing Together Geographical and Sociological Imaginations*, ed. J. Agnew and J. Duncan (Boston: Unwin Hyman, 1989), 30.
34. Ibid. 41

35. Clifford Geertz, *The Interpretation of Cultures* (New York: Basic Books, 1973), 261.
36. John I. H. Baur, *The Inlander: Life and Work of Charles Burchfield, 1893–1967* (Newark: Associated University Presses, 1982), 169.
37. "About Your Region," United States Economic Development Administration, accessed July 15, 2014, www.knowyourregion.org/about
38. James Howard Kunstler, "Clusterfuck Nation: A Glimpse into the Future," accessed July 15, 2014, www.kunstler.com/mags_ure.htm
39. Thomas Alfred Jackson, "Evolution by Revolution," *The Socialist Standard*, May 1905.
40. Umberto Eco, "Reports from the Global Village" in *Travels in Hyperreality* (San Diego: Harcourt Brace Jovanovich, 1986), 142–143.
41. Mikhail Bakhtin, *Rabelais and His World*, trans. H. Iswolsky (Bloomington: Indiana University Press, 1964), 83.
42. Albert Camus, *The Rebel: An Essay on Man in Revolt* (New York: Alfred A. Knopf, 1956), 296.
43. Hesiod, *Theogony/Works and Days*, trans. M. West (Oxford: Oxford University Press, 1988), 9.
44. Yi-Fu Tuan, "Place: An Experiential Perspective." *The Geographical Review*, Vol. 65, No. 15 (April 1975): 153.
45. Krishan Kumar, "Utopian Thought and Communal Practice: Robert Owen and Owenite Communities." *Theory and Society*, Vol. 19, No.1 (Feb. 1990): 21.
46. Edwin Starr, "Time," *War & Peace* (Detroit: Motown Records, 1970).

Bibliography

"About Your Region." United States Economic Development Administration. Accessed July 15, 2014. www.knowyourregion.org/about

"Bloody Interesting Decision." *Rocky Mountain News*. February 7, 1990. Accessed June 5, 2014. www.coloradoaim.org/history/19900207rmnopedrussellmeanssentence.htm

"Bush on Security Efforts." *Washington Post*. Sept 12, 2001. Accessed June 5, 2014. www.washingtonpost.com/wp-srv/nation/transcripts/bushtext_091201.html

"CCTV Cameras (CYCLOPS)." *Power Telecomm, Inc.* Accessed May 12, 2014. www.powertelecomm.com/product/cctv.html

"DSM-5 Diagnostic Criteria for PTSD Released." United States Department of Veterans Affairs, National Center for PTSD. December 5, 2012. Accessed June 5, 2014. www.ptsd.va.gov/professional/pages/diagnostic_criteria_dsm-5.asp

"End of Tyburn." *The Glasgow Herald*. September 28, 1964.

"Groundbreaking Study Finds U.S. Security Industry to be $350 Billion Market; Integrators Vital to End Users' Decision-Making." *SDM Magazine*. August 20, 2013. Accessed June 5, 2014. www.sdmmag.com/articles/89562-groundbreaking-study-finds-us-security-industry-to-be-350-billion-market-integrators-vital-to-end-users-decision-making

"'Guilty as Hell' Pleads Cop Killer 'Bucky' Phillips." *Police One*. November 30, 2006. Accessed June 5, 2014. www.policeone.com/legal/articles/1194278-Guilty-as-hell-pleads-cop-killer-Bucky-Phillips/

"Haymarket Monument Unveiled." *The New York Times*. May 21, 1889.

"Haymarket Statue Removed." *The New York Times*. July 15, 1900.

"Holidays, Time off, Sick Leave, Maternity and Paternity Leave." *Working, Jobs and Pensions*. Accessed May 27, 2014. https://www.gov.uk/browse/working

"Influence and Lobbying: Agribusiness." Center for Responsive Politics. Accessed June 2, 2014. www.opensecrets.org/industries/indus.php

"Lulz." *Encyclopedia Dramatica*. Accessed October 15, 2013. https://encyclopediadramatica.es/Lulz

"Memorial Placed on Site Of Tyburn Tree." *San Francisco Call*. Vol. 105, No. 173. May 22, 1909.

"Monument on the Move." *Chicago History*. Accessed June 5, 2014, www.chicagohistory.org/dramas/epilogue/toServeAndProtect/monumentOnTheMove.htm

"Occupy Olympus at NY International Fringe Festival 2013." Magis Theatre Company. Accessed June 24, 2014. http://magistheatre.org/current_production.html

"Pinsent-Masons Launches 2010–11 Water Industry 'Bible.'" Pinsent-Masons. November 10, 2010. Accessed June 2, 2014. www.pinsentmasons.com/en/media/press-releases/2010/pinsent-masons-launches-2010–11-water-industry-bible/

"Post Traumatic Stress Disorder." Nebraska Department of Veterans Affairs. 2007. Accessed June 5, 2014. www.ptsd.ne.gov/what-is-ptsd.html

"Protest Banners Hung From Acropolis." *Al Jazeera*. December 17, 2008. www.aljazeera.com/news/europe/2008/12/2008121718556743523.html

"The Tyburn." *Wetherspoon*. Accessed June 5, 2014. www.jdwetherspoon.co.uk/home/pubs/the-tyburn

"Welcome to Marble Arch." City of Westminster. Accessed October 30, 2013. www.westminster.gov.uk/myparks/parks/marble-arch/

"Yes, A Police Riot." *The New York Times*. August 26, 1988.

Aberley, Douglas. *Boundaries of Home: Mapping for Local Empowerment*. New York: New Society Publishers, 1992.

Abler, Ronald, John S. Adams, and Peter Gould. *Spatial Organization*. Englewood Cliffs: Prentice Hall, 1971.

Abousnnouga, Gil and David Machin. "War Monuments and the Changing Discourses of Nation and Soldiery." In *Semiotic Landscapes: Language, Image, Space*, edited by Adam Jaworski and Crispin Thurlow, 219–240. London and New York: Continuum International, 2010.

Adelman, William J. "The True Story behind the Haymarket Police Statue." In *Haymarket Scrapbook*, edited by David Roediger and Franklin Rosemont 167–170. Chicago: Charles H. Kerr, 1986.

Alinksy, Saul. *Rules for Radicals*. New York: Vintage Books, 1971.

Allen, John. *Rethinking the Region: Spaces of Neoliberalism*. New York: Routledge, 1998.

Allison, Robert J. *The Boston Massacre*. Beverly: Commonwealth Editions, 2006.

Andersen, Nels. *The Milk and Honey Route: A Handbook for Hobos*. New York: Vanguard Press, 1931.

Anderson, Benedict. *Imagined Communities: Reflections on the Origin and Spread of Nationalism*. London: Verso, 1991.

Anderson, Perry. *Lineages of the Absolutist State*. London: Verso, 1974.

Anonymous. *United Mission Statement*. Accessed October 15, 2013. http://anoncentral.tumblr.com/post/11964602984/the-text-of-the-well-received-flier-being-passed-around

Aristotle. *Politics*. Translated by Benjamin Jowett. Oxford: Clarendon Press, 1905.

Atgeld, John Peter. *The Pardon of the Haymarket Prisoners*. June 26, 1893. Accessed June 5, 2014. http://law2.umkc.edu/faculty/projects/ftrials/haymarket/pardon.html#REASONS_FOR_PARDONING

Athaneaus. *Deipnosophistae*. New York: Loeb Classical Library, 1929.

Bakhtin, Mikhail. *Rabelais and His World*. Translated by Helene Iswolsky. Bloomington: Indiana University Press, 1964.

Barber, Benjamin. *Jihad vs. McWorld: Terrorism's Challenge to Democracy*. London: Corgi, 2003.

Baudrillard, Jean. *Simulacra and Simulation*. Translated by Sheila Faria Glaser. Ann Arbor: University of Michigan Press, 1994.

Baur, John I. H. *The Inlander: Life and Work of Charles Burchfield, 1893–1967*. Newark: Associated University Presses, 1982.

Bayly, C. A. *The Birth of the Modern World, 1780–1914*. Malden: Blackwell, 2004.

Bede, The Venerable. *The Reckoning of Time*. Translated by Fait Wallis. Liverpool: Liverpool University Press, 1999.

Bentham, Jeremy. *The Panopticon Writings*. Edited by Miran Bozovic. London: Verso, 1995.

Bierce, Ambrose. *The Unabridged Devils Dictionary*. Edited by David E. Schultz and S. J. Joshi. Athens: University of Georgia Press, 2000.

Biggs, Michael. "Putting the State on the Map: Cartography, Territory and European State Formation." *Comparative Studies in Society and History*, Vol. 41, No. 2 (April 1999): 374–405.

Binneveld, Hans. *From Shellshock to Combat Stress: A Comparative History of Military Psychology*. Amsterdam: Amsterdam University Press, 1997.

Bishop, Julia C. and Mavis Curtis. *Play Today in the Primary School Playground*. Buckingham: Open University Press, 2001.
Blair, W. Granger. "Newcastle Rises From the Doldroms: Once Backward English City is Being Transformed." *The New York Times*. February 27, 1966.
Blakely, Edward James and Mary Gail Snyder. *Fortress America: Gated Communities in the United States*. Washington: The Brookings Institution, 1997.
Borges, Jorge Luis. "The Aleph." In *The Aleph and Other Short Stories*, edited and translated by Norman Thomas di Giovanni 15–30. London: Jonathan Cape, 1970.
Borst, Arno. *The Ordering of Time: From the Ancient Computus to the Modern Computer*. Oxford: Polity Press, 1993.
Bradby, Barbara. "The Destruction of the Natural Economy." *Economy and Society*, Vol. 4, No. 2 (1975): 127–161.
Brautigan, Richard. *All Watched Over By Machines of Loving Grace*. San Francisco: The Communication Company, 1967.
Breisach, Ernst. *Historiography: Ancient, Medieval and Modern, Third Edition*. Chicago: University of Chicago Press, 2007.
Brennan Center for Justice, Liberty and National Security Program. "How 9/11 Changed the Law." *Mother Jones*, September 9, 2011.
Brod, Craig. "Managing Technostress: Optimizing the Use of Computer Technology." *Personnel Journal*, Vol. 61, No. 10 (October 1982): 753–757.
Brooke, Alan and David Brandon. *Tyburn: London's Fatal Tree*. Gloucestershire: Sutton, 2004.
Brown, Emma. "D.C. Parents Push For More Recess." *The Washington Post*. August 30, 2013. Accessed June 5, 2014. http://articles.washingtonpost.com/2013-08-30/local/41600378_1_montgomery-county-schools-recess-minimum
Bruni, Leonardo. *History of the Florentine Peoples*. Translated by James Hankins. Boston: Harvard University Press, 2001.
Bush, George W. "Address to a Joint Session of Congress and the American People," *Office of the Press Secretary*. September 20, 2001.
Byrne, Edmund F. *Public Power, Private Interests and Where to We Fit In?* New York: 1st Book Library, 1998.
Caldwell, Simon. "Council Plans New Memorial to Tyburn Martyrs." *The Catholic Harold*. January 30, 2009. Accessed June 5, 2014.http://archive.catholicherald.co.uk/article/30th-january-2009/2/council-plans-new-memorial-to-the-tyburn-martyrs
Campbell, Joseph. *The Hero with A Thousand Faces*. New York: New World Library, 2008.
Camus, Albert. *The Rebel: An Essay on Man in Revolt*. New York: Alfred A. Knopf, 1956.
Carr, E. H. *What Is History?* New York: Penguin Books, 2008.
Cassiodorus. *The Letters of Cassiodorus*. Translated by Thomas Hodgkin. London: Henry Frowde, 1886.
Charnovitz, Steve. "Addressing Environmental and Labor Issues in the World Trade Organization." *Trade and Global Markets: World Trade Organization*. Progressive Policy Institute. November 1, 1999. Accessed June 5, 2014. www.dlc.org/ndol_cid308.html?kaid=108&subid=128&contentid=649
Chinn, Carl. *Poverty amidst Prosperity: The Urban Poor in England, 1834–1914*. Manchester: Manchester University Press, 1995.
City of Gary Indiana. "Dollar Home Project." Accessed May 14, 2014. www.teamgaryindiana.com/?p=1232
Ciuraru, Carmela. *Nom De Plume, A (Secret) History of Pseudonyms*. New York: Harper Collins, 2011.
Claasen, Jo Marie. *Displaced Persons: The Literature of Exile from Cicero to Boethius*. Madison: University of Wisconsin Press, 1999.

Clifford, James. "Traveling Cultures." In *Cultural Studies*, edited by Lawrence Grossberg, Cary Nelson, and Paula Treicher, 96–111. New York: Routledge, 1992.
Close, Colonel Sir Charles. *The Early Years of the Ordnance Survey*. Devon: David and Charles Reprints, 1969.
Cooper, James Fenimore. *The Spy: A Tale of Neutral Ground*. New York: W.A. Townsend and Company, 1861.
Coote, Anna, Andrew Sims, and Jane Franklin. *21 Hours*. New Economics Foundation. February 13, 2010. Accessed June 5, 2015. www.neweconomics.org/publications/entry/21-hours
Cornwell, Janelle. "Worker Cooperatives and Spaces of Possibility: An Investigation of Subject Space at Collective Copies." *Antipode*, Vol. 44, No. 3, (2012): 725–44.
Cowley, Julian. "Free Your Mind, Your Ass Will Follow." *The Wire*. No. 295, September 2008: 36–40.
Cronon, William. *Changes in the Land: Indians, Colonists and the Ecology of New England*. New York: Hill and Wang, 1983.
Cummins, Chip. "Efforts to Secure Oil Pipeline Linking Iraq and Turkey Fail." *Wall Street Journal*. December 13, 2003. Accessed June 15, 2014. http://online.wsj.com/article/SB106867930345823400.html
Dagne, Ted. "Somalia: Conditions and Prospects for Lasting Peace." *CRS Report for Congress*. Washington: Congressional Research Service, August 13, 2011.
Davies, Lizzie. "Occupy Movement: City-By-City Police Crackdowns So Far." *The Guardian*. November 15, 2011. Accessed June 15, 2014. www.theguardian.com/world/blog/2011/nov/15/occupy-movement-police-crackdowns?CMP=twt_gu
Davis, Angela. "Masked Racism: Reflections on the Prison Industrial Complex." *Color Lines*. Sept 10, 1998. Accessed June 15, 2014. http://colorlines.com/archives/1998/09/masked_racism_reflections_on_the_prison_industrial_complex.html
Davis, J. C. *Utopia and the Ideal Society: A Study of English Utopian Writing 1516–1700*. Cambridge: Cambridge University Press, 1981.
Debord, Guy. *The Society of the Spectacle*. Canberra: Treason Press, 2002.
DeCasseres, Benjamin. "The Hit and Run Thinker: Technocracy Analyzed and Frisked." *Life Magazine*. Feb. 1933. 100.
Deman, Paul. "Literary History and Literary Modernity." In: *Blindness and Insight: Essays in the Rhetoric of Contemporary Criticism*, 142–165. London: Methuen, 1983.
Derrida, Jacques. *Specters of Marx*. New York: Routledge, 1994.
Derrida, Jacques. *The Beast and the Sovereign, Vol.1*. Chicago, University of Chicago, 2009.
Develin, R. "'Provocatio' and Plebiscites. Early Roman Legislation and the Historical Tradition." *Mnemosyne*. Fourth Series, Vol. 31, Fasc. 1 (1978): 45–60.
Dickey, Colin. "Unhousing." *The Paris Review*. Last modified March 19, 2014. Accessed June 15, 2015. www.theparisreview.org/blog/2014/03/19/unhousing/
Dixon, Eddie. "Willie McCool Memorial." *City of Lubbock, Department of Parks and Recreation*. May 7, 2005. Accessed May 14, 2014. www.mylubbock.us/departmental-websites/departments/parks-recreation/top-navigation-menu-items/parks/attractions/willie-mccool-memorial
Dobson, R. B. *The Peasants' Revolt of 1381*. London: Macmillan Press, 1970.
D'Oca, Daniel. "The Arsenal of Inclusion and Exclusion." *MAS Context*. No. 17 (Spring 2013): 54–75.
Dorn Van-Rossum, Gerhard. *The History of the Hour: Clocks and Modern Temporal Orders*. Translated by Thomas Dunlap. Chicago and London: University of Chicago Press, 1996.
Drengson, Alan R. "Shifting Paradigms: From Technocrat to Planetary Persons." In *The Deep Ecology Movement: An Introductory Anthology*, edited by Alan R. Drengson and Yuichi Inoue, 74–100. Berkeley: North Atlantic Books, 1995.

Duck, Stephen. "The Thresher's Labour." In *Poems for Several Occasions*, 7–16. London: John Osborn, 1738.
Dunn, Alastair. *The Great Rising of 1381: The Peasants' Revolt and England's Failed Revolution*. Stroud: Tempus Press, 2002.
Eco, Umberto. "Reports from the Global Village." In: *Travels in Hyperreality*. San Diego: Harcourt Brace Jovanovich, 1986.
Edmunds, R. David. *Tecumseh: The Quest for Indian Leadership*. New York: Pearson Longman, 2007.
Eksteins, Modris. *Rites of Spring: The Great War and the Birth of the Modern Age*. New York: Vintage Press, 1989.
Ellement, John R. "An Upgrade of History." *Boston Globe*. May 11, 2011. Accessed June 5, 2014. www.boston.com/news/local/massachusetts/articles/2011/05/11/upgrade_to_raise_boston_massacre_sites_visual_impact/
Emanuele, Vince. "Interview with David Harvey: Rebel Cities and Urban Resistance Part II." *Z Magazine*. January 7, 2013. Accessed June 5, 2014. www.zcommunications.org/contents/190562
Emerson, Ralph Waldo. "Uses of Great Men." In *Representative Men: Seven Lectures*. 1–21. Boston: Houghton Mifflin and Company, 1888.
Engels, Friedrich. *The Condition of the Working-Class in England in 1844*. London: Swan Sonnenschein & Co., 1892.
Entrikin, J. Nicholas. "Place, Region and Modernity." In *The Power of Place: Bringing Together Geographical and Sociological Imaginations*, edited by John S. Agnew and James S. Duncan, 30–43. Boston: Unwin Hyman, 1989.
Esarey, Logan, ed. *Messages and Letters of William Henry Harrison*. Indianapolis: Indiana Historical Commission, 1922.
Everdell, William. *The First Moderns: Profiles in the Origins of Twentieth Century Thought*. Chicago: University of Chicago Press, 1997.
Exodus 20:8–11. *The Bible: Authorized King James Version*. Edited by Robert Carroll and Stephen Prickett. Oxford: Oxford University Press, 2008.
Field Manual No. 3–24: *Counterinsurgency*. Washington, D.C. Department of the Army, 2006.
Fischer, Frank. *Technocracy and the Politics of Expertise*. New York: Sage, 1990.
Fitz, Karsten. "Commemorating Crispus Attucks: Visual Memory and the Representations of the Boston Massacre, 1770–1857." *Amerikastudien*, Vol. 50, No. 3 (2005): 463–484.
Fogelson, Robert M. *Downtown: Its Rise and Fall 1880–1950*. New Haven: Yale University Press, 2001.
Forbes, Eric G. *Greenwich Observatory: One of Three Volumes by Different Authors Telling the Story of Britain's Oldest Scientific Institution*. London: Taylor & Francis, 1975.
Foucault, Michel. *Discipline and Punish: The Birth of the Prison*. New York: Penguin, 1977.
Fox, Charles M. *Psychopathology of Hysteria*.Boston: The Gorham Press, 1913.
Francese, Christopher. *Ancient Rome in So Many Words*. New York: Hippocrene, 2007.
Franklin, Benjamin. "The Morals of Chess." In *The Chess Player*, edited by George Walker Teacher, 7–12. Boston: N. Dearborn, 1841.
Friedman, Jonathan. "Being in the World: Globalization and Localization." In *Global Culture: Nationalism, Globalization and Modernity*, edited by Mike Featherstone, 311–328. London: Sage Publications, 1990.
Garrison, Lloyd. "Nigeria Regains Business Capital." *The New York Times*. May 15, 1966.
Gay, Peter. *The Enlightenment: The Science of Freedom*. New York: Norton, 1977.
Geertz, Clifford. *The Interpretation of Cultures*. New York: Basic Books, 1973.
Gellner, Ernest. *Thought and Change*. London: Weidenfield and Nicolson, 1964.

Giddens, Anthony. *A Contemporary Critique of Historical Materialism*. Berkeley and Los Angeles: University of California Press, 1983.
Giddens, Anthony. *Modernity and Self Identity: Self and Society in the Late Modern Age*. Stanford: Stanford University Press, 1991.
Giddens, Anthony. *The Consequences of Modernity*. Cambridge: Polity Press, 1990.
Goody, Jack. *The Theft of History*. Cambridge: Cambridge University Press, 2006.
Gordon, Robert J. "Is U.S. Economic Growth Over? Faltering Innovation Confronts The Six Headwinds." *NBER Working Paper No. 18315*. Last modified August 2012. Accessed June 5, 2014. www.nber.org/papers/w18315
Gordon, Robert J. "Why Innovation Won't Save Us." *Wall Street Journal*. Dec. 21, 2012.
Gottleib, Robert. *Environmentalism Unbound: Exploring New Pathways For Change*. Boston: MIT Press, 2002.
Graves, Robert. *The Greek Myths: Vol. 1*. New York: Penguin Books, 1955.
Grossman, Jonathan. "Fair Labor Standards Act of 1938: Maximum Struggle for a Minimum Wage." *Monthly Labor Review*, Vol. 101, No. 6 (June 1978): 22–30.
Grossman, Reinhardt. *The Existence of the World: An Introduction to Ontology*. London: Routledge, 1992.
Guinness, Henry Grattan. *The Divine Programme of the World's History*. New York: Hodder and Stoughton, 1888.
Hannity, Sean. "Barack Obama, Community Organizing Years." *Hannity's America Video Archive*, Fox News Corporation. April 29, 2011. Accessed June 5, 2014. www.foxnews.com/hannitysamerica/
Harley, J. Brian. "Deconstructing the Map." *Cartographica*, Vol. 26, No. 2 (Spring 1989): 1–20.
Harris, Ed and Frank Sammartino. "Trends in the Distribution of Household Income, 1979–2009." Congressional Budget Office. August 6, 2012. Accessed June 5, 2014. www.cbo.gov/sites/default/files/cbofiles/attachments/Trends_in_household_income_forposting.pdf
Harvey, David. *A Brief History of Neoliberalism*. Oxford: Oxford University Press, 2005.
Harvey, David. *The Condition of Postmodernity: An Enquiry into the Origins of Cultural Change*. Cambridge: Blackwell, 1989.
Harvey, P.D.A. *Mappa Mundi, The Hereford World Map*. Toronto: University of Toronto Press, 1996.
Harvey, Rowland Hill. *Robert Owen: Social Idealist*. Los Angeles: University of California Press, 1949.
Hays, Constance. "Home Videos Turn Lenses on the Police: Police Find Their Actions Tracked by Amateurs With Videocameras." *New York Times*. August 15, 1988.
Hegel, Georg Wilhelm Friedrich. *Hegel: The Essential Writings*. Edited by Frederick G. Weis. New York: Harper & Row, 1974.
Hegel, Georg Wilhelm Friedrich. *Lectures on the Philosophy of History*. London: G. Bell and Sons, 1910.
Hegel, Georg Wilhelm Friedrich. *The Phenomenology of Spirit*. Translated by A. V. Miller. Oxford: Clarendon Press, 1977.
Hegel, Georg Wilhelm Friedrich. *The Philosophy of History*. Translated by J. Sibree and C. J. Friedrich. New York: Dover, 1956.
Hegel, Georg Wilhelm Friedrich. *The Philosophy of Right*. Translated by S. W. Dyde. Kitchener: Batoche Books, 2001.
Heidegger, Martin. *Being and Time*. Translated by Joan Stambaugh. Albany: State University of New York Press, 1996.
Henley, Will. "What Value Should We Place on Water in Developing Countries?" *The Guardian*. September 17, 2013. Accessed June 5, 2014. www.theguardian.com/sustainable-business/value-water-developing-countries-talkpoint

Hesiod. *Theogony/Works and Days*. Translated by M. L. West. Oxford: Oxford University Press, 1988.
Hesiod. *Hesiod and Theogonis*. Translated by Dorothea Wender. New York: Penguin Books, 1973.
Hilton, R. H. "Small Town Society in England before The Black Death." In *The Medieval Town 1200–1540*, ed. Richard Holt and Gervase Rosser, 71–97. New York: Longman, 1990.
Hilton, Rodney. *Class Conflict and the Crisis of Feudalism: Essays in Medieval Social History*. London: Hambledon Press, 1985.
Hindle, Brooke. *Emulation and Invention*. New York: NYU Press, 1981.
Hirschkorn, Phil. "New York Lays Cornerstone for Freedom Tower." *CNN*. July 6, 2004. Accessed June 5, 2014. http://edition.cnn.com/2004/US/Northeast/07/04/wtc.cornerstone/index.html
Hobsbawm, Eric. *Nations and Nationalism since 1780*. Cambridge: Cambridge University Press, 1997.
Hobsbawm, Eric. "Mass Producing Traditions: Europe, 1870–1914." In *The Invention of Tradition*, edited by Eric Hobsbawm and Terence Ranger, 70–5. Cambridge: Cambridge University Press, 1996.
Hobsbawm, Eric. *Primitive Rebels, Studies in Archaic Forms of Social Movement in the 19th and 20th Centuries*. Manchester: Manchester University Press, 1959.
Hodgkin, Katherine and Susannah Radstone. "Patterning the National Past." In *Contested Pasts: The Politics of Memory*, edited by Katherine Hodgkin and Susannah Radstone, 169–174. London and New York: Routledge, 2003.
Hooker, Clarence. *Life in the Shadows of the Crystal Palace, 1910–27: Ford Workers in the Model T Era*. Bowling Green: Bowling Green State University Popular Press, 1997.
Horsman, Reginald. "The Dimensions of an 'Empire of Liberty': Expansion and Republicanism, 1775–1825." *Journal of the Early Republic*, Vol. 9, No. 1 (Spring 1989): 1–30.
Howard, John. *The State of the Prisons in England and Wales with Preliminary Observations, and an Account of Some Foreign Prisons*. Warrington: William Eyres, 1777.
Howard, John. "An Account of the Principal Lazerettos of Europe." In: *The Works of John Howard, Esq. Vol. II*. London: Johnson, Dilly and Cadell, 1791.
Hughes, C. J. "Bohemia with a Softer Edge: Tompkins Square Park." *New York Times*. October 31, 2004.
Husari, Ruba "Oil Assessment Work Gets Going in North Iraq." *International Oil Daily*. May 2, 2003. Accessed June 5, 2014. www.iraqoilforum.com/?tag=kirkuk-ceyhan
Jackson, Peter. *Maps of Meaning*. London: Taylor & Francis, 1989.
Jackson, Thomas Alfred. "Evolution by Revolution." *The Socialist Standard*. May 1905.
Jacobs, Jane. *The Death and Life of Great American Cities*. New York: Modern Library, 1961.
Jameson, Frederic. *A Singular Modernity: Essay on the Ontology of the Present*. London: Verso, 2002.
Jefferson, Thomas. "Memoir of Meriweather Lewis." In *The History of the Lewis and Clark Expedition*, edited by Elliott Coues, Vol. 1, xv–xxxiii. New York: Francis P. Harper, 1893.
Jennings, Francis. "Virgin Land and Savage People." *American Quarterly*, Vol. 23, No. 4 (October 1971): 519–541.
Johnson, Alvin W. and Frank H. Yost. *Separation of Church and State in the United States*. Minneapolis: University of Minnesota, 1948.
Jones, Dan. *Summer of Blood: The Peasants' Revolt of 1381*. London: Harper, 2010.

Kallen, Horace. "Democracy versus the Melting Pot: A Study of American Nationality." *The Nation.* Feb. 25, 1915.

Kant, Immanuel. "Idea for a Universal History from a Cosmopolitan Point of View." In *Immanuel Kant, On History*, translated by Lewis White Beck 11–27. New York: Bobbs-Merrill Co., 1963.

Kant, Immanuel. "Prolegomena to Any Future Metaphysics." In *Classics of Western Philosophy*, edited by Steven M. Cahn 876–1033. Indianapolis: Hackett Publishing, 1990.

Kazin, Michael and Steven J. Ross. "America's Labor Day: The Dilemma of a Workers' Celebration." *The Journal of American History*, Vol. 78, No. 4 (Mar. 1992): 1294–1393.

Keim, Brandon. "Nanosecond Trading Could Make Markets Go Haywire." *Wired Magazine.* February 16, 2012. Accessed June 5, 2014. www.wired.com/wiredscience/2012/02/high-speed-trading/

Kelly, Gordon P. *A History of Exile in the Roman Republic.* Cambridge: Cambridge University Press.

Kennedy, Paul. *The Rise and Fall of Great Powers.* New York: Vintage Books, 1987.

Kifner, John. "New York Closes Park to Homeless." *New York Times.* June 4, 1991.

Kipling, Rudyard. "The White Man's Burden." In *Kipling Stories and Poems Every Child Should Know*, edited by W. T. Chapin, 359–361. Cambridge: Riverside Press, 1909.

Klein, Naomi. *The Shock Doctrine.* New York: Penguin, 2008.

Knight, Stephen. *Robin Hood: A Complete Study of the English Outlaw.* Oxford: Blackwell Publishing, 1994.

Kumar, Krishan. "Utopian Thought and Communal Practice: Robert Owen and Owenite Communities." *Theory and Society*, Vol. 19, No.1 (Feb. 1990): 1–35.

Kundera, Milan. *Slowness: A Novel.* Translated by Linda Asher. New York: Harper Collins, 1997.

Kunstler, James Howard. "Clusterfuck Nation: A Glimpse into the Future." Accessed July 15, 2014. www.kunstler.com/mags_ure.htm

Kunstler, James Howard. "Eyesore of the Month: July 2005." Accessed June 5, 2014. www.kunstler.com/eyesore_200507.html

Kunstler, James Howard. *The Geography of Nowhere: The Rise and Decline of America's Man Made Landscape.* New York: Free Press, 1994.

Kurzweil, Ray. "The Law of Accelerating Returns." March 7, 2001. Accessed July 15, 2014. www.kurzweilai.net/the-law-of-accelerating-returns

Kusmer, Kenneth L. *Down and Out, the Homeless in American History.* Oxford: Oxford University Press, 2002.

Lampert, Evgeni. *The Apocalypse of History.* London: Faber and Faber, 1948.

Langdon-Davies, John. "Ramie: King Cotton's Rival." *The Forum*, Vol. 90 (November 1933): 289–292.

Larcom, Lucy. *A New England Girlhood, Outlined from Memory.* New York: Houghton and Mifflin, 1889.

Laubach, Rev. Frank. *Why There Are Vagrants: A Study.* New York: Columbia, 1916.

Le Corbusier. *The Modulor.* Translated by Peter de Francia and Anna Bostock. Basel: Birkhauser Architecture, 2000.

Lee, Jennifer. "EPA Said to Avoid Studies Conflicting White House." *New York Times.* July 14, 2003.

LeFebvre, Henri. *The Production of Space.* Translated by Donald Nicholson-Smith. Malden and Oxford: Blackwell Publishing, 1991.

Le Goff, Jacques. *Time, Work and Culture in the Middle Ages.* Chicago: University of Chicago Press, 1980.

Leopold, Aldo. "Engineering and Conservation." In *The River of the Mother God, and Other Essays by Aldo Leopold*, edited by Susan L. Flader and J. Baird Callicott, 249–254. Madison: University of Wisconsin Press, 1991.
Letters, Samantha. *Gazetteer of Markets and Fairs in England and Wales to 1516*. London: Centre for Metropolitan Research, 2004. Last modified Dec. 16, 2013, www.history.ac.uk/cmh/gaz/gazweb2.html
Levine, David. "Recombinant Family Formation Strategies." *Journal of Historical Sociology*, Vol. 2, No. 2 (1989): 89–115.
Levinson, Sanford. *Written in Stone: Public Monuments in Changing Societies*. Durham: Duke University Press, 1998.
Lewis, Martin W. and Karen E. Wigen. *The Myth of Continents: A Critique of Metageography*. Berkeley, Los Angeles, London: University of California Press, 1997.
Lind, Michael. "The Age of Turboparaylsis." *Salon*. Dec. 27, 2011. www.salon.com/2011/12/27/the_age_of_turboparalysis/
Linebaugh, Peter and Marcus Rediker. *The Many Headed Hydra: Sailors, Slaves, Commoners and The Hidden History of the Revolutionary Atlantic*. Boston: Beacon Press, 1999.
Linebaugh, Peter. *The Magna Carta Manifesto: Liberty and Commons for All*. Berkeley: University of California, 2008.
Llobera, Josep. *The God of Modernity: The Development of Nationalism in Western Europe*. Oxford: Berg, 1994.
MacMullen, Ramsay. "Market-Days in the Roman Empire." *Phoenix*, Vol. 24, No. 4 (1970): 333–341.
Macrobius. *Saturnalia*. Boston: Loeb Classical Library, 2011.
Martin, Douglas. "Motley Crowd Calls Park Special Place." *New York Times*. July 9, 1999.
Marx, Karl. *Capital: A Critical Analysis of Capitalist Production*. London: Swan & Sonnenschein & Co., 1902.
Marx, Karl and Friedrich Engels. *The Communist Manifesto*. 1848. Reprint. New York: Bedford/St. Martins, 1999.
Marx, Karl. "Theses on Feuerbach." In *The Viking Portable Marx*, edited by Eugene Kamenka, 155–157. New York: Viking Penguin, 1983.
Mathews, Shailer. *The Spiritual Interpretation of History*. Cambridge: Harvard University Press, 1920.
Maxwell, Steve. *A Concise Review of Challenges and Opportunities in the World Water Market*. 2012. Accessed July 15, 2014. www.summitglobal.com/documents/Maxwell2012WaterMarketReview-a030912.pdf
McClintock, Harry. "The Big Rock Candy Mountains." Victor #21704, 1928. Accessed July 15, 2014. http://archive.org/details/TheBigRockCandyMountains
McCord, David J. (ed.)*The Statutes at Large of South Carolina. Vol. 7, Containing the Acts Relating to Charleston, Courts, Slaves, and Rivers*. Columbia: A.S. Johnston, 1840.
McDowell, Douglas M. *The Law in Classical Athens*. London: Thames and Hudson, 1978.
McElvaine, Robert. *The Great Depression: America 1929–1941*. New York: Crown Publishing, 2009.
McFadden, Robert. "Park Curfew Protest Erupts into a Battle." *New York Times*. August 8, 1988.
McIntyre, Douglas. "24 Hour Stock Trading." *24/7 Wall Street*. February 15, 2013. Accessed July 15, 2014. http://247wallst.com/investing/2013/02/15/24-hour-stock-trading/
McKenna, Terence. *The Invisible Landscape: Mind, Hallucinogens and the I-Ching*. New York: Seabury, 1975.

McNamee, Tom. "After 118 Years, Haymarket Memorial to Be Unveiled." *Chicago Sun Times*. September 7, 2004.
Meyer III, Stephen. *The Five Dollar Day: Labor Management and Social Control in the Ford Motor Company, 1908–21*. Albany: State University of New York Press, 1981.
Michel, Lou and Dan Herbeck. "Phillips Admits Killing State Trooper Ex-Fugitive Also Pleads Guilty to Wounding Two Other Troopers." *The Buffalo News*. November 20, 2006.
Milton, John "Paradise Lost." In *John Milton: Complete Poems and Major Prose*, edited by Merritt Y. Hughes, 173–469. New York: Macmillan, 1957.
Minkowski, Hermann. "Raum und Zeit." *Physikalische Zeitschrift*, Vol. 10, No. 104 (1909): 75–88.
Mommsen, Theodore. "Petrarch's Conception of 'The Dark Ages." *Speculum*, Vol. 17, No. 2 (April 1942): 226–242.
Moore, Gordon E. "Cramming More Components onto Integrated Circuits." *Electronics* Vol. 38, No. 8 (April 19, 1965): 114–117.
Moore, Sylvia. "Occupy Wall Street Movement on Health Care." California One Care. October 15th, 2011. Accessed July 15, 2014. http://californiaonecare.org/occupy-wall-street-movement-on-health-care/
Morales, Helen. *Classical Mythology, a Very Short Introduction*. Oxford: Oxford University Press, 2007.
More, Sir Thomas. *Utopia*. London: Richard Chiswell, 1684.
Motesharrei, Safa, Jorge Rivas, and Eugenia Kalnay. "A Minimal Model for Human and Nature Interaction." November 13, 2012. Accessed. July 15, 2014. www.ara.cat/societat/handy-paper-for-submission-2_ARAFIL20140317_0003.pdf
Moya, Jose C. "The Massification of International Families in the Nineteenth Century." In *Transregional and Transnational Families in Europe and Beyond: Experiences since the Middle Ages*, edited by Christopher Johnson, 23–42. New York: Berghahn Books, 2011.
Mumford, Lewis. "The Neighborhood and Neighborhood Unit." *The Town Planning Review*, Vol. 24, No. 4 (January 1954): 256–270.
Mumford, Lewis. *The Pentagon of Power*. New York: Harcourt Brace Jovanovich, 1970.
Myers, Charles S. "A Contribution to the Study of Shell Shock." *The Lancet*. February 13, 1915.
Nadkarni, Maya. "The Death of Socialism and the Afterlife of its Monuments." In *Contested Pasts: The Politics of Memory*, edited by Katherine Hodgkin and Susannah Radstone, 193–207. London and New York: Routledge, 2003.
Nagle, D. Brendan and Stanley M. Burstein. *The Ancient World: Readings in Social and Cultural History*. Englewood Cliffs, NJ: Prentice Hall, 1995.
Nasdaq Regular Trading Session Schedule. Accessed May 27, 2014, www.nasdaq.com/about/trading-schedule.aspx
Newburgh, William of. *Historia Rerum Anglicarum*. Edited by Scott McLetchie. London: Seeley's, 1861.
Nicholas, David. *The Growth of the Medieval City, from Late Antiquity to the Early Fourteenth Century*. Harlow: Addison Wesley Longman, 1997.
Niebuhr, Reinhold. "The Tower of Babel." In: *Reinhold Neibuhr: Theologian of Public Life*, edited by Larry Rasmussen 82–87. Minneapolis: First Fortress, 1991.
Nieves, Evelyn. "Tensions Remain at Closed Tompkins Square Park." *New York Times*. June 17, 1991.
Nobles, Gregory. "Straight Lines and Stability: Mapping the Political Order of the Anglo-American Frontier." *The Journal of American History*, Vol. 80, No. 1 (June 1993): 9–35.

Ost, Laura. "NIST Ytterbium Atomic Clock Sets Record for Stability." *NIST Tech Beat*. August 22, 2013. Accessed July 15, 2015. www.nist.gov/pml/div688/clock-082213.cfm

Parenti, Christian. *The Soft Cage: Surveillance in America, From Slavery to the War on Terror*. New York: Basic Books, 2004.

Pausanias. *Description of Greece with an English Translation*. Translated by W.H.S. Jones and H.A. Ormerod. London: William Heinemann Ltd., 1918.

Pendergrast, Christopher. "Codeword Modernity." *New Left Review* Vol. 24, (November-December 2003): 95–111.

Perlman, Fredy. *Against His-Story, Against Leviathan!* Detroit: Black & Red, 1983.

Petersen, Kevin. "Room To Grow: Detroit Takes First Steps to Legalize Urban Agriculture." *Michigan Journal of Environmental and Administrative Law*. February 8, 2013. Accessed July 15, 2014. http://students.law.umich.edu/mjeal/2013/02/room-to-grow-detroit-takes-the-first-steps-to-legalize-urban-agriculture/

Pickering, Danby (ed.)*The Statutes at Large, From Magna Chart to the End of the Eleventh Parliament of Great Britain*. London: Joseph Bentham, 1764.

Pierce, Charles Sanders. *The New Elements of Mathematics*, edited by Carolyn Eisele. Atlantic Highlands: Humanities Press, 1976.

Plato. *The Sophist and the Statesman*. Translated by A. E. Taylor. London: Thomas Nelson and Sons, 1961.

Plato. *Timaeus*. Translated by Benjamin Jowett. Rockville: Serenity, 2009.

Pliny. *The Natural History of Pliny*. Translated by John Bostock and H.T. Riley. London: Henry G. Bohn, 1855.

Plutarch. *Aristedes*. Translated by Bernadotte Perrin. New York: Charles Scribner's Sons, 1901.

Polanyi, Karl. "Ports of Trade in Early Societies." In: *Primitive, Archaic and Modern Economies*, edited by George Dalton. 238–260. New York: Doubleday, 1968.

Pollan, Michael. *Food Rules: An Eater's Manual*. New York: Penguin Books, 2009.

Pournelle, Jerry. "The View from Chaos Manor." *View* 406. April 3–9, 2006. Accessed July 15, 2015. www.jerrypournelle.com/archives2/archives2view/view408.html#Iron

Price, George. *A Treatise on Fire and Thief Proof Depositories and Locks and Keys*. London: Simpkin, Marshall and Co., 1856.

Punter, David. *Modernity*. New York: Palgrave Macmillan, 2007.

Purdum, Todd. "Melee in Tompkins Sq. Park: Violence and its Provocations." *New York Times*. August 14, 1988.

Rainbow, Paul. "Space Knowledge and Power." In *The Foucault Reader*, edited by Paul Rainbow 239–256. New York: Pantheon Books, 1984.

Rauch, Erik. *Productivity and the Workweek*. Accessed May 28, 2014. http://groups.csail.mit.edu/mac/users/rauch/worktime/

Rediker, Marcus. *Between the Devil and the Deep Blue Sea: Merchant Seamen, Pirates, and the Anglo-American Maritime World, 1700–50*. Cambridge: Cambridge University Press, 1987.

Reece, Florence. "Which Side Are You on?" *Here Comes A Wind: Labor on the Move*, Vol. 4, 1–2: 90. Chapel Hill: Institute for Southern Studies, 1976.

Reid, Douglas. "The Decline of Saint Monday 1766–1876." *Past and Present*, Vol. 71, No. 1 (May 1976): 76–101.

Richter, Daniel. *Facing East from Indian Country*. Cambridge: Harvard University Press, 2003.

Roos, Jerome. "Exarchia and the Greek Spirit of Resistance." *Roarmag*. July 18, 2011. Accessed July 15, 2015. http://roarmag.org/2011/07/exarchia-and-the-greek-spirit-of-resistance/

Rushkoff, Douglas. *Present Shock: When Everything Happens Now.* New York: Current Press, 2013.
Saint Augustine. *Confessions.* Translated by J. G. Pilkington. Edinburgh: T.T. & Clark, 1876.
Saulny, Susan. "The Blackout: Sense of Community." *New York Times.* August 16, 2003.
Saussere, Ferdinand de. *Course in General Linguistics.* New York: Columbia University Press, 2011.
Schensul, Jean J., Margaret D. LeCompte, Robert T. Trotter II, Ellen K. Cromley, and Merrill Singer. *Mapping Social Networks, Spatial Data and Hidden Population.* Walnut Creek: Altamira Press, 1999.
Schmookler, Andrew Bard. *The Illusion of Choice: How the Market Economy Shapes Our Destiny.* Albany: State University of New York Press, 1993.
Schneider, John C. *Detroit and the Problem of Order.* Lincoln: University of Nebraska Press, 1980.
Schweiterman, Joseph P. and Dana M. Caspall. *The Politics of Place: A History of Zoning in Chicago.* Chicago: Lake Claremont Press, 2006.
Scott, Howard. *Technocracy: Science vs. Chaos.* Akron: Technocracy Inc., 1933.
Seeger, Charles. "On Dissonant Counterpoint." *Modern Music,* Vol. 7, No.4 (June-July 1930): 25–31.
Segal, Howard P. *Technological Utopianism in American Culture.* Syracuse: Syracuse University Press, 2005.
Semple, Janet. *Bentham's Prison: A Study of the Panopticon Penitentiary.* Oxford: Oxford University Press, 1993.
Shackel, Paul A. "Remembering Haymarket and the Control for Public Memory." In *Heritage, Labor and the Working Classes,* edited by Laurijane Smith, Paul Shackel, and Gary Campbell, 34–51. Oxon: Routledge Press, 2011.
Shalev, Eran. "Ancient Masks, American Fathers: Classical Pseudonyms during the American Revolution and Early Republic." *Journal of the Early Republic,* Vol. 23, No. 2 (Summer 2003): 151–172.
Sharp, Granville. *The Gilbart Prize Essay on the Adaption of Recent Discoveries and Inventions in Science and Art to the Purposes of Practical Banking.* London: Groombridge and Sons, 1854.
Shatan, Chaim F. "Post-Vietnam Syndrome." *New York Times.* May 6, 1972.
Shatter, Rev. Carlos. *Sir Humfrey Glylberte and His Enterprise of Colonization in America.* Boston: Publications of the Prince Society, 1903.
Sheridan, Richard B. *Sugar and Slavery: An Economic History of the British West Indies, 1623–1775.* Jamaica: Canoe Press, 1974.
Smith, John. *Travels and Works.* Edited by Edward Arber. Edinburgh: John Grant, 1910.
Smith, Thomas Carlyle. *Native Sources of Japanese Industrialization 1750–1920.* Berkeley: University of California Press, 1988.
Smyth, William Henry. "Technocracy—National Industrial Management." In *Technocracy.* 7–15. Berkeley: W.H. Smyth, 1920.
Snyder, Gary. *The Real Work: Interviews and Talks, 1964–1979.* New York: New Directions Books, 1980.
Solnit, Rebecca. *Wanderlust: A History of Walking.* New York: Verso Books, 2002.
Sopko, Kate. *Stewards of Lost Lands.* Cleveland: The Language Foundry, 2008.
Spier, Fred. *Big History and the Future of Humanity.* Malden: Wiley-Blackwell, 2011.
Starr, Edwin. "Time." In *War & Peace.* Detroit: Motown Records, 1970.
Staw, Barry M. "Knee Deep in the Big Muddy: A Study of Escalating Commitment to a Chosen Course of Action." *Organizational Behavior and Human Performance,* Vol. 16, Issue 1 (1976): 27–44.

Stearns, Peter N. "A Cease Fire for History." *The History Teacher*, Vol. 30, No. 1, (November 1996): 65–81.
Stearns, Peter N. *The Industrial Revolution in World History*. Philadelphia: Perseus, 2013.
Sugden, John. "Tecumseh's Travels Revisited." *Indiana Magazine of History*, Vol. 96, No. 2 (June 2000): 150–168.
Sulzberger, C. L. "Europe Asks Questions on U.S. Foreign Policy." *The New York Times*. May 11, 1952.
Tames, Richard L., ed. *Documents of the Industrial Revolution 1750–1850*. London: Hutchinson, 1971.
The Invisible Committee. *The Coming Insurrection*. Los Angeles: Semiotext(e), 2009.
The Royal Academy. "The Eighty-Seventh Exhibition, 1855." *The Art-Journal*. June 1, 1855.
Thompson, E. P. "Time, Work-Discipline and Industrial Capitalism." *Past and Present*, No. 28 (December 1967): 56–97.
Thompson, E. P. *Whigs and Hunters: The Origin of The Black Act*. London: Penguin Books, 1975.
Thompson, Mark. "The $5 Trillion War on Terror." *Time Magazine*. June 29, 2011.
Thrower, Norman J. *Maps and Civilization: Cartography in Culture and Society*. Chicago: University of Chicago Press, 2008.
Toffler, Alvin. *Future Shock*. New York: Random House, 1970.
Tompkins Square Park Temperance Fountain. *City of New York Parks and Recreation*. Accessed June 5, 2014. www.nycgovparks.org/parks/tompkinssquarepark/monuments/1558
Tonnies, Ferdinand. *Community and Civil Society*. Edited by Jose Harris. Cambridge: Cambridge University Press, 2001.
Toynbee, Arnold. *Experiences*. Oxford: Oxford University Press, 1969.
Tuan, Yi-Fu. "Place: An Experiential Perspective." *The Geographical Review*, Vol. 65, No. 15 (April 1975): 151–165.
Tuan, Yi-Fu. *Space and Place: An Experiential Perspective*. Minneapolis: University of Minnesota Press, 1977.
Tucker, Glen. *Tecumseh: Vision of Glory*. New York: Cosimo, 2005.
Tucker, T. G. *Life in Ancient Athens*. London: Macmillan, 1907.
Turner, Frederick Jackson. *The Frontier in American History*. New York: Frederick Holt & Co., 1920.
Ullman, Harlan K. and James P. Wade Jr. *Rapid Dominance—A Force for All Seasons*. London: Royal United Services Institute For Defence Studies, 1998.
Underwood, G. and R.A. Swain, "Selectivity of Attention and the Perception of Duration." *Perception*, Vol. 2, No. 1 (1973): 101–105.
United Nations Population Fund. "Urbanization: A Majority in Cities." Last modified May, 2007. Accessed July 15, 2015. www.unfpa.org/pds/urbanization.htm
Van Der Veer, Peter. "The Global History of 'Modernity." *Journal of Economic and Social History of the Orient*. Vol. 41, No. 3 (1998): 285–294.
Veblen, Thorstein. *The Engineers and the Price System*. New York: B.W. Heubsch, 1921.
Vey, Tristan. "Welcome to the Anthropocene, Earth's New Era." *Worldcrunch*. Last modified October 26, 2011. Accessed July 15, 2015. www.worldcrunch.com/welcome-anthropocene-earth-s-new-era/tech-science/welcome-to-anthropocene-earth-s-new-era/c4s3982/
Vico, Giambattista. *The New Science of Giambattista Vico*. Translated by Thomas Goddard Bergin and Max Harold Fisch. Ithaca: Cornell University Press, 1948.
Walsh, Gerald. *Industrialization and Society: Selected Sources*. London: McClelland and Steward, 1969.

Warf, Barney. *Time-Space Compression: Historical Geographies*. New York, Routledge: 2008.
Webster, T.B.L. *Everyday Life in Classical Athens*. New York: G.P. Putnam, 1969.
White, Hayden. *Metahistory: The Historical Imagination of Nineteenth Century Europe*. Baltimore: Johns Hopkins University Press, 1973.
Whitman, Alden. "Mumford Finds City Strangled by Excess of Cars and People." *The New York Times*. March 22, 1967.
Williamson, Arthur. *Apocalypse Then: Prophecy and the Making of the Modern World*. Westport: Praeger Publishers, 2008.
Wines, Michael. "Behind the Park Melee, a New Generation Gap." *New York Times*. August 8, 1988.
Wines, Michael. "Class Struggle Erupts Along Avenue B." *New York Times*. August 10, 1988.
Wilson, Michael. "Dragnet Yields Whimsy and Dread Upstate." *New York Times*. July 15, 2006.
Wing, Charles. *The Evils of the Factory System, Demonstrated by Parliamentary Evidence*. London: Frank Cass & Co., 1967.
Winthrop, John. *The Journal of John Winthrop: 1630–1649*. Edited by Richard Dunn and Laetitia Yeandle. Boston: Massachusetts Historical Society, 1996.
Wittfogel, Karl August. *Oriental Despotism: A Comparative Study of Total Power*. New Haven and London: Yale University Press, 1957.
Wohlgelernter, Maurice. *Israel Zangwill: A Study*. New York: Columbia University Press, 1964.
Wordie, J.R. "Chronology of English Enclosure, 1500–1914." *The Economic History Review*, Vol. 36, No. 4 (November 1983): 483–505.
Worster, Donald. *Rivers of Empire: Water, Aridity and the Growth of the American West*. New York and Oxford: Oxford University Press, 1985.
York, Byron. "What Did Obama Do As a Community Organizer?" *The National Review*. Sept. 8, 2008.
Zangwill, Israel. *The Melting Pot*. London: William Heineman, 1914.
Zerzan, John. "Why Primitivism?" 2002. Accessed July 15, 2014. www.johnzerzan.net/articles/why-primitivism.html
Zobel, Hiller B. *The Boston Massacre*. New York: W.W. Norton & Company, 1970.

Index

Aberley, Douglas 169
Acropolis 138
Ages of Man 1, 5, 13, 15
Alexandria 20
Alfred, King 66
Alinsky, Saul 175
Allen, John 177
Althingi 20
American Indian Movement 138
American Revolution 7, 41, 75–6, 104, 125, 141, 180
Ananke 2, 5, 42
anarcho-primitivism 37
ancien regime 17
Anderson, Benedict 59–60, 184
Anderson, Nels 112
Anderson, Perry 97
Anonymous 74, 76–7
anthropocene 24
Apocalypse 7; Apollo spacecraft 54; as final judgement 14, 125; in historical thought 18, 41, 127; Mayan 38; as resolution 24
Arab Spring 9, 55, 177
Aristophenes 75
Aristotle 25, 28–9, 65, 129
Assange, Julian 70
Athelsane 66
Athens 20, 69, 129, 138
Augustulus 18

Bakhtin, Mikhail 182
Barber, Benjamin 86
Baudrillard, Jean 37
Bayly, C. A. 22
Bede, The Venerable 14–15
Belial 132
Bentham, Jeremy 84–6
Beuth, Christian P. W. 94

Bierce, Ambrose 63
Biggs, Michael 63, 178
Bin Laden, Osama 152
Birmingham 36, 60, 108
Blake, William 41
Blakely, Edward J. 109
Bloomberg, Michael 154
blue laws 66–7
Boccaccio, Giovanni 18
Bonarparte, Napoleon 105, 180
Booth, Charles 57
Borges, Jorge Luis 55–6, 58
Boston Massacre 140–2, 144
Brabeck-Letmathe, Peter 88
Bradby, Barbara 99
Brautigan, Richard 53
Bretton Woods 70
Brod, Craig 95
Brogger, Mary 149–151
Bruni, Leonardo 18, 22
Burchfield, Charles 178–9
Bush, George W. 153

Campbell, Joseph 14
Camus, Albert 183–4
Canadian Shield 50–1
carnivalesque 182
Carr, E. H. 21
Carthage 20
Cass, Lewis 146
Cassini Survey 63
Cassiodorus, Magnus Aurelius 17–18
Cellarius, Christoph 18, 22
Chaos 1–5; as antithesis of order 15, 24–5, 40, 67, 91, 123, 128–9, 136, 170; in mythology 28, 50, 107, 160, 162
Charles II 58, 66, 103
Cheney, Dick 158

chess 27
Christmas Truce 60
Ciuraru, Carmela 76
Clear Hold Build 9, 124, 154–60, 162
Clear Skies Initiative 87
Cleisthenes 129
Clifford, James 82–3
clock time 61–72
Cogswell, Henry 162
Cold War 38, 54, 91, 148
Columbus, Christopher 20, 138, 169
commons 35, 112–13, 131–4, 145, 156, 176–7
communism 1, 37, 91, 93, 125, 176
Communist Manifesto 34, 37
Congleton 66
Constantine 66
Cooper, James Fenimore 77
cooperatives 174, 176, 185
Cornwell, Jane 174
Crusades 69
Crutzen, Paul 24
Cyclopes 1–5; in mythology 29, 41, 50–1, 71, 75, 107; and surveillance 129, 136, 160, 169, 172, 176, 181, 184–5

Dahomey 68, 93
Darcy, Thomas 161
Da-sein 51
Das Kapital 64, 87
Davis, Angela 83
Debord, Guy 114, 124–5, 127–8, 139
DeCassares, Benjamin 90
declinist theory 38
Deep Ecology Movement 98
de Lacy, Henry 66
DeMan, Paul 124
democracy 7; as form of resistance 133–4, 138; and modernity 90, 93, 100, 116, 154; as universal construct 20–1
Derrida, Jacques 35–6, 43, 128, 182
de Tabbeleye, William 66
Detroit 64, 116, 185
Dewitt, Clinton C. 116
Dickey, Colin 36
Diderot, Denis 19
D'Oca, Daniel 109–10
Donahue, Mark 151
Drengson, Alan R. 83
Duck, Stephen 67
dynastic realm 63, 178

Eco, Umberto 182
Edward I 66
Edward VI 111
Einstein, Albert 22
Elizabeth I 111, 156, 158
Ellington, Duke 27
Emerson, Ralph Waldo 172, 178, 181, 185
emporia 65, 69
enclosures 35; enforcement of 57, 110; and industrialization 60, 63, 87, 101; resistance to 131, 145, 159, 176
Engels, Friedrich 9, 33–5, 39, 50, 94
Enlightenment 21–2, 25, 29, 97, 125, 138
Entrikin, J. Nicholas 178
Eratosthenes 56, 103
Erobos 28
eschatology 15, 37–42, 127, 185
Eustis, William 145
Everdell, William 23
Ewing, William G. 145
Exodus, Book of 174

Fair Labor Standards Act 72
Farmer, Doyne 67
Fawkes, Guy 77
Fischer, Frank 93
Fixnetix 67
Fogelson, Robert M. 69
Ford Sociological Department 116
Fort Meigs 140, 144–7
Fort Miamis 145
Foucault, Michel 10, 85, 88, 169
Franklin, Benjamin 27
Freedom Tower 153–4
Friedman, Jonathan 99
Fringe Festival 75
Fry, Joshua 104–5

Garrison, Lloyd 93
gated communities 109
Gay, Peter 21–2
Geertz, Clifford 178
Geist 31–2, 35
Gellner, Ernest 59
Genesis, Book of 14
gentrification 159–162
Geographical Information Systems (GIS) 57, 170
Giddens, Anthony 23, 40, 42–3, 50, 99
Gilbert, Sir Humphrey 157–8

globalization 6; debated meaning 27; and modernity 16, 20, 70, 74, 86, 99, 172; resistance to 44, 114, 174
Golden Torch 186
Goody, Jack 22
Gordon, Robert 38
Goya, Francisco 125–6
Gray, Thomas 36
Great Depression 90–1, 177
Great Zimbabwe 20
Green Revolution 9, 177
Grossman, Reinhardt 28
Guinness, Henry Gratton 39
Gulf of Aden 134
Guthrie, Woody 74

Halliburton 158
Han Dynasty 56
Hannity, Sean 175
Harlan County 42
Harley, J. Brian 169, 171
Harness, Henry Drury 57
Harris, Jonathan 58
Harrison, William Henry 145–6
Harvey, David 43, 88, 99, 113
Haudenosaunee 20
Hauntology 35–7
Haymarket Affair 140, 147–51
Hebe 160
Hegel, Georg Wilhelm Friedrich 5; and concept of final causes 7, 39, 41, 127; historical theories of 31–5, 50–1
Heidegger, Martin 51
Hekatonkheires 1–5; as metaphor for modernity 41, 50, 59, 75, 129, 136, 172; in mythology 29, 71
Henry III 66
Henry VI 66
Henry VIII 110
Hereford Map 56
Heron, Gil Scott 181
Hesiod 1–5; and "Ages of Man" 13–15, 18, 28, 35, 65, 113, 125, 145, 169, 177; and Titanomachy 50, 68, 123, 128, 184, 186
Hilton, R. H. 69
Hindle, Brooke 53
hobos 111–13
Hobsbawm, Eric 59–60, 132
Hodgkin, Katherine 137
Holocene 24

Howard, John 84
Humanism 18, 52
Huntington, Samuel 38
hydraulic civilization 100–1

I Ching 40
Illinois Labor History Society 149
Indignados 75
Industrial Revolution 6; effects on biosphere 24; effects on economy 96, 109; effects on historical analysis 31, 87; effects on labor 60–1, 63, 67, 71, 94, 125
Information Age 7, 87
International Monetary Fund (IMF) 70

Jackson, Peter 172
Jackson, Thomas Alfred 180, 183
Jacobite Rebellion 57
Jacobs, Jane 173
James I 66
Jameson, Frederic 29
Jamestown 157
Jean D'Alembert 50
Jefferson, Peter 104–5
Jefferson, Thomas 104–5
Jennings, Francis 156

Kahokia 20
Kali-Yuga 14
Kallen, Horace 116–17
Kandahar 69
Kant, Immanuel 25, 29–33, 50
Kazin, Michael 73
Kennedy, John F. 38
Kennedy, Paul 38
Kiguba 20
Kipling, Rudyard 171
Kirkuk-Ceyhan Oil Pipeline 158
Klein, Naomi 100
Knutsford 66
Koch, Edward 161
Kronos 1–5; association with "Golden Age" 13, 28–9, 50–1, 113, 125, 169, 177; as symbol of resistance 75, 107, 129, 136, 185–6; and time 42
Kundera, Milan 54
Kunstler, James Howard 102, 153–4, 180
Kurzweil, Ray 23

Labor Day 73
Lagash 62

206 *Index*

Lampert, Evgeni 127
Land of Cockaigne 113
Land Ordinance of 1785 63, 104
Langdon-Davies, John 90
Larcom, Lucy 94
latifundia 100–1
Laubach, Frank 111
Le Corbusier 76
Le Goff, Jacques 61, 6
Leeds 60
Lefebvre, Henri 51, 68–71, 138–9, 170, 181
Leopold, Aldo 55
Levine, David 86
Levinson, Sanford 137
Lewis, Martin W. 26
Lewis, Meriweather 105
Libeskind, David 154
Lind, Michael 22
Linneaus, Carolus 19
Llobera, Josep 59
Loeb, Harold 91
Lord Ashley's Mines Commission 94
Louisiana Purchase 105–6
Lowell, Massachusetts 94
Lynch, Bill 162

McClintock, Harry 112–3
McCool, William C. 59
McIntyre, Douglas 67
McKenna, Terence 40
Macrobius 66
Magna Carta 176
Mammon 132
Manchester 60, 94
Manning, Chelsea 70
Marble Arch 144
Marquis, Samuel S. 116
Marx, Karl 33–7; and economics 64, 86, 100, 111, 174; and politics 169; and teleology 41, 127
Massachusetts Bay Colony 38
Mathews, Shailer 25
Maunder, E. Walter 58
Means, Russell 138
metageography 26
metahistory 26
metatheory 7, 24–7
Milton, John 132
Minkowski, Hermann 50
Moloch 132
monoculture 8–11; agricultural 96–8, 100–3; as effect of globalization 20, 103–18, 169–72, 175–6; social 98–9, 128–9, 133–4, 162, 164, 181–5; and technocracy 61, 74, 77–8, 82–9, 123–5, 178–9
Moore, Gordon E. 23
Moore, Sylvia 75
Moore's Law 23, 137
Morales, Helen 3
More, Thomas 110
Morton's Fork 86
Moya, Jose 23
Mumford, Lewis 152, 172–3
Myers, Charles 95

Nadkarni, Maya 139
Nagatsune, Okura 53
NASA Goddard Institute for Space Studies 117
National Institute of Standards and Technology 53
Nationalism 6–7, 19–21, 51–2, 58–71, 77–8, 82, 89, 99, 106, 115, 117, 127, 133, 152–3, 181, 184
National Regeneration Society 72
Negro Act of 1740 136
Neibuhr, Richard 137
Neighborhood 44; as place 64, 110, 114, 159–62, 172–4, 177–9, 185–6; as biblical concept 174–5
Nemesis 184
Newcastle-Upon-Tyne 93
New Deal 91
New Economics Foundation 73
Newton, Isaac 11
Northern Soul Movement 186
nundinae 65–6

Obama, Barack 175
Occupy Movement 9, 44, 74–6, 177
Oldmixon, John 98
Ontology 7–8, 27–9, 162
Ordnance Survey 57, 63
Oroboros 103
ostracism 129–30, 136
Owen, Robert 72

Packer, George 38
Palmyra 68
Panopticon 84–5, 88
Parenti, Christian 70, 159
Pausanias 1
Peasants' Revolt of 1381 130–1
Pendergrast, Christopher 29
Perlman, Fredy 82
Perrysburg 140, 145–6

Petrarch 18–9, 22
Pherecrates 113
Phillips, Ralph "Bucky" 134
Pierce, Charles Sanders 3
piracy 15, 132–4
Plato 10, 25, 27–9, 42, 103
Pliny the Elder 101
Plutarch 129
Polanyi, Karl 68
Polo, Marco 20
Pompeius 36
Port Au Prince 20
Post Traumatic Stress Disorder (PTSD) 96
postmodernity 5, 17, 43, 50, 99, 149
Power Telecomm 2
Price, George 107–8
Primitive Accumulation 24, 97
prison-industrial complex 83–6
Procter, Henry 145–6
Proetus 5
Putnam, Robert 177
Putney 20

Radstone, Susannah 137
Ramsden Theodolite 57
Rauch, Erik 72
Read, Isabella 94
Reagan, Ronald 39, 93, 109, 117, 160
Recording Industry of America (RIAA) 134
Rediker, Marcus 133
Reece, Florence 42
Reformation 41
region 9, 44, 65, 69, 140, 153, 172, 176–80, 184–6
Regionalist Art Movement 178
Reid, Douglas 71–2
Renaissance 6, 18, 21, 52, 63, 76
Revelation, Book of 14, 39, 56, 125
Revere, Paul 140
Richard II 130–1
Riot Act of 1715 135–6
Robin Hood 133–5
Rome 16–18, 20, 36, 75, 100–1, 109, 115, 130, 136
Roosevelt, Theodore 115
Ross, Steven J. 73
Royal Greenwich Observatory 58, 103
Rushkoff, Douglas 24
Rust Belt 177, 184

Saint Augustine 49
Saint John 14

Saint Monday 71–2
Satan 132
Saussere, Ferdinand 3
Scaliger, Joseph Justus 52–3
Schleiffen plan 54
Scholasticism 18, 25
Scott, Howard 90–1
Scotus, John Duns 25
Seattle 9, 44, 177, 185
Secessio Plebis 75
Seeckt, Hans Van 54
semiotics 2–4
Semple, Janet 88
September 11, 2001 74, 140, 152–4
Seven Years War 104, 145
Shalev, Eran 76
Shapter, Thomas 57
Sharp, Granville 108
Shatan, Chaim 96
Sheffield 60
shell shock 95
Sheridan, Richard B. 97–8
Shock and Awe 54
Sibree, J. 32
Singer, Merrill 171
Situationists 124–5
Slow Food Movement 44, 73
Smyth, William Henry 89–90, 93, 96
Snow, John 57
Snowden, Edward 70
Snyder, Gary 98
Snyder, Mary Gail 109
Solnit, Rebecca 55
Sons of Liberty 77
Sopko, Kate 164
South Sea Economic Collapse 76
Starr, Edwin 186
Staw, Barry M. 118
Stearns, Peter 17
Stoke-on-Trent 186
Stono Rebellion 136
Sulzberger, C. L. 91
Swain, Rodney A. 55

Tal Afar 158
tamkarum 20
Tamralipta 20
Tartarus 1; as metaphor 38, 71, 160; in mythology 3, 29, 75, 128, 132
Taylor, Frederick Winslow 95
TechKNOWLEDGEy Strategic Group 88
Technocracy 83, 89–96, 117, 184
Technostress 95
Tecumseh 144–6

208 Index

Tehrir Square 7, 20, 177, 180, 185
teleology 7–11; and eschatology 37–42; as historical theory 29–32, 35, 51, 78, 100, 127, 139, 162, 185
Tenochtitlan 20
The Aleph 55–6, 58
The Art-Journal 89, 93, 95–6
The Black Act 77
The Coming Insurrection 82, 113–4, 117, 172, 176
The Golden Age 1; as metaphor 15, 133; in mythology 4, 5, 13, 33, 50, 145, 177
The Great War 19, and industrialization 89, 94, 54, 124; and nationalism 59, 153–4
The Invisible Committee 82, 89, 117, 176
The Melting Pot 115–17
Thomas-Rasset, Jammie 134
Thompson, E. P. 61, 67
Tiananmen Square 7, 20
time-space compression 16, 43–4, 99
time-space distanciation 43, 99
Time Wave Zero 40
Titanomachy 1–4; as metaphor 10, 42, 68, 169, 184, 186; in mythology 33, 51, 123, 128
Titus 101
Toffler, Alvin 23–4, 124
Tompkins Square Park 160–2
Tonnies, Ferdinand 176
Tuan, Yi-Fu 21, 40, 139, 185
Tucker, T. G. 65
turboparalysis 22
Turner, Frederick Jackson 106
Turner, Nat 70
Tyburn 140–4
Tyler, Wat 131
Typhon 5, 50, 169

Ullman, Harlan K. 54
Umma 62

Underwood, Geoffrey 55
U. S. Economic Development Administration (EDA) 179

vagabondage 110–11
Van Der Veer, Peter 22
Veblen, Thorsten 90
Vico, Giambattista 42
Virilio, Paul 43

Wade, James P. Jr. 54
War of 1812 144–6
Warf, Barney 43
Washington Monument 139
Western Civilization 16–17, 21, 37, 50, 60, 109
White, Hayden 26
Whydah 68
Wigen, Karen E. 26
William of Newburgh 141
William of Ockham 25
Williamson, Arthur 41
Winthrop, Jonathan 38
Wittfogel, Karl 100–1
Works and Days 13, 65, 113
World Bank 70
World Trade Center 152–3
World War I *see* The Great War
Worster, Donald 101
Wright Brothers 54

Xi'an 20

Yeats, William Butler 41
York, Byron 175

Zangwill, Israel 115
Zerzan, John 37, 181
Zeus 1–5; as metaphor 37–8, 50–1, 75, 91, 129, 136, 160, 169; in mythology 13, 29, 33, 42, 123;
Zheng He 20
Zucotti Park 74–5

For Product Safety Concerns and Information please contact our EU representative GPSR@taylorandfrancis.com
Taylor & Francis Verlag GmbH, Kaufingerstraße 24, 80331 München, Germany

www.ingramcontent.com/pod-product-compliance
Lightning Source LLC
Chambersburg PA
CBHW061444300426
44114CB00014B/1830